Toxicological Relevance of EDCs and Pharmaceuticals in Drinking Water

Awwa
Research
Foundation

About the Awwa Research Foundation

The Awwa Research Foundation (AwwaRF) is a member-supported, international, nonprofit organization that sponsors research to enable water utilities, public health agencies, and other professionals to provide safe and affordable drinking water to consumers.

The Foundation's mission is to advance the science of water to improve the quality of life. To achieve this mission, the Foundation sponsors studies on all aspects of drinking water, including supply and resources, treatment, monitoring and analysis, distribution, management, and health effects. Funding for research is provided primarily by subscription payments from approximately 1,000 utilities, consulting firms, and manufacturers in North America and abroad. Additional funding comes from collaborative partnerships with other national and international organizations, allowing for resources to be leveraged, expertise to be shared, and broad-based knowledge to be developed and disseminated. Government funding serves as a third source of research dollars.

From its headquarters in Denver, Colorado, the Foundation's staff directs and supports the efforts of more than 800 volunteers who serve on the board of trustees and various committees. These volunteers represent many facets of the water industry, and contribute their expertise to select and monitor research studies that benefit the entire drinking water community.

The results of research are disseminated through a number of channels, including reports, the Web site, conferences, and periodicals.

For subscribers, the Foundation serves as a cooperative program in which water suppliers unite to pool their resources. By applying Foundation research findings, these water suppliers can save substantial costs and stay on the leading edge of drinking water science and technology. Since its inception, AwwaRF has supplied the water community with more than $300 million in applied research.

More information about the Foundation and how to become a subscriber is available on the Web at **www.awwarf.org.**

Toxicological Relevance of EDCs and Pharmaceuticals in Drinking Water

Prepared by:
Shane A. Snyder and **Rebecca A. Trenholm**
Southern Nevada Water Authority
Applied Research and Development Center, Henderson, NV 89015

Erin M. Snyder, **Gretchen M. Bruce**, and **Richard C. Pleus**
Intertox, Inc., Seattle, WA 98121

and

Jocelyn D.C. Hemming
Wisconsin State Laboratory of Hygiene
2601 Agriculture Drive, Madison, WI 53718

Jointly sponsored by:
Awwa Research Foundation
6666 West Quincy Avenue, Denver, CO 80235-3098

WateReuse Foundation

California Urban Water Agencies

and

Tailored Collaboration partners:
Southern Nevada Water Authority and other co-funding utilities

Published by:

Awwa Research Foundation

Distributed by:

American Water Works Association

IWA Publishing

DISCLAIMER

This study was jointly funded by the Awwa Research Foundation (AwwaRF), WateReuse Foundation (WRF), California Urban Water Agencies (CUWA), Southern Nevada Water Authority (SNWA), and the other co-funding utilities. AwwaRF, WRF, CUWA, SNWA, and the other utilities assume no responsibility for the content of the research study reported in this publication or for the opinions or statements of fact expressed in the report. The mention of trade names for commercial products does not represent or imply the approval or endorsement of AwwaRF, WRF, CUWA, SNWA, or the other utilities. This report is presented solely for informational purposes.

TABLE OF CONTENTS

LIST OF TABLES

LIST OF FIGURES

FOREWORD

The Awwa Research Foundation is a nonprofit corporation that is dedicated to the implementation of a research effort to help utilities respond to regulatory requirements and traditional high-priority concerns of the industry. The research agenda is developed through a process of consultation with subscribers and drinking water professionals. Under the umbrella of a Strategic Research Plan, the Research Advisory Council prioritizes the suggested projects based upon current and future needs, applicability, and past work; the Collaborative Research, Research Application, and Tailored Collaborations programs; and various joint research efforts with organizations such as the U.S. Environmental Protection Agency, the U.S. Bureau of Reclamation, and the Association of California Water Agencies.

This publication is a result of one of these sponsored studies, and it is hoped that its findings will be applied in communities throughout the world. The following report serves not only as a means of communicating the results of the water industry's centralized research program but also as a tool to enlist further support of the nonmember utilities and individuals.

Projects are managed closely from their inception to the final report by the foundation's staff and a large cadre of volunteers who willingly contribute their time and expertise. The foundation serves a planning and management function and awards contracts to other institutions such as water utilities, universities, and engineering firms. The funding for this research effort comes primarily from the Subscription Program, through which water utilities subscribe to the research program and make an annual payment proportionate to the volume of water they deliver and consultants and manufacturers subscribe based upon their annual billings. The program offers a cost-effective and fair method for funding research in the public interest.

A broad spectrum of water supply issues is addressed by the foundation's research agenda: resources, treatment and operations, distribution and storage, water quality and analysis, toxicology, economics, and management. The ultimate purpose of the coordinated effort is to assist water suppliers to provide the highest possible quality of water economically and reliably.

The true benefits are realized when the results are implemented at the utility level. The foundation's trustees are pleased to offer this publication as a contribution toward that end.

David E. Rager
Chair, Board of Trustees
Awwa Research Foundation

Robert C. Renner, P.E.
Executive Director
Awwa Research Foundation

David [illegible]

Robert G. Renner, P.E.

ACKNOWLEDGMENTS

This project would not have been possible without the collaboration and team work of a vast network of water experts in a diversity of locations across the United States.

To this end, the research team would like to acknowledge the AwwaRF project manager for this study, Dr. Djanette Khiari who worked relentlessly to keep this project on track and provide expert insight and guidance through every aspect of this effort. We also wish to thank our colleagues at the Applied Research and Development Center (ARDC) at the Southern Nevada Water Authority (SNWA) including Linda Parker, Spencer Porter, Oscar Quiñones, Dave Rexing, Ben Stanford, Brett Vanderford, Ron Zegers and Janie Zeigler. Our team is indebted to Roger Bueher and JC Davis from SNWA Public Information who led the project team's efforts on public communication and outreach. We also want to thank the SNWA executive team, including Kay Brothers, Thomas Minwegen, Patricia Mulroy and Richard Wimmer, without whose support and encouragement the ARDC would not exist. We greatly appreciate the assistance of the SNWA Information Systems Department, particularly Collin Burnell, Joe Garza, Frankie Lewis, and Jeff Waufle. Dr. Hongxia (Dawn) Lei was a post-doctoral researcher at SNWA and contributed during the initial phases of the project. The Intertox team would like to thank their colleagues Erin Bennet, Ammon Gilbert, Susan Leaman and Michael Peterson.

The research team would also like to share our sincere appreciation of the global efforts related to trace contaminants in drinking water. Without question we recognize the research that has been conducted, and continues to be developed, around the world. We could not possibly review and cite each and every publication and report that has been developed on this emerging topic, but we clearly recognize our colleagues who have provided guidance and inspiration for our project.

Again, our team strongly encourages continued collaborative efforts among the water industry, the regulatory community, and the public. These covalent bonds bring together a diverse group of people who all seek one common goal, protection of public health. These valuable efforts should not be discouraged should we continue to pursue proactive harvesting of knowledge for the greater good of all mankind.

EXECUTIVE SUMMARY

INTRODUCTION

Analytical technology has made it possible to detect and quantify nearly any compound known to humankind at diminishingly minute concentrations in water. Although the earliest reports of pharmaceuticals and steroid hormones in water date back nearly four decades, it is only within the past decade that the subject has come to the forefront of scientific and public attention. Today we know definitively that trace levels of pharmaceuticals, potential endocrine disrupting compounds (EDCs), and other emerging contaminants occur in source water, and to a lesser extent, in finished drinking water. Based on research thus far, it appears that many conventional and advanced treatment processes will greatly reduce the concentrations of these compounds. Nevertheless, considering the continued advancements in analytical technologies, today's non-detectable contaminants will be tomorrow's emerging contaminants. If presence/absence becomes our litmus test for risk and subsequent actions, treatment technology will be increasingly, and perhaps unnecessarily, costly and energy intensive. This is an especially important consideration due to the energy cost and greenhouse gas emissions of advanced treatment. For these reasons, it is of utmost importance to determine human health-based screening levels from which meaningful treatment goals and analytical detection limits can be established. This study provides critical information regarding the occurrence and risk assessment for pharmaceuticals and potential EDCs in drinking water. The information in this report will allow utilities and regulators to begin to make informed decisions regarding the relevance of trace pharmaceuticals and EDCs.

STUDY OBJECTIVES

1. Determine a representative group of indicator compounds based on toxicity, treatability, occurrence, structure, and analytical capabilities.

2. Develop robust analytical protocols for selected indicator compounds that provide accurate and precise identification and quantification in drinking water and recycle water matrices.

3. Determine the occurrence of selected pharmaceuticals and potential EDCs in a diverse set of U.S. drinking waters.

4. Evaluate the strengths and weaknesses of using an *in vitro* bioassay for measuring estrogenicity in water and various food/beverage items.

5. Conduct risk analysis for representative pharmaceuticals and potential EDCs.

6. Develop communication tools related to pharmaceuticals and EDCs in drinking water.

PROJECT APPROACH

The initial phase of this study involved a thorough review of published literature and databases to determine a representative group of indicator compounds. Of the thousands of potential candidates, 62 chemicals were chosen for further evaluation based upon likelihood of occurrence, production volume, toxicity, and analytical capability. This suite of chemicals included 20 pharmaceuticals (and active metabolites), 26 potential EDCs, five steroid hormones, and 11 phytoestrogens (natural estrogens from plants). Robust analytical methods using both liquid chromatography with tandem mass spectrometry (LC-MS/MS) and gas chromatography with tandem mass spectrometry (GC-MS/MS) were developed specifically for this project. The resulting method reporting limits (MRL) ranged from 0.2 to 120 nanograms per liter (ng/L, or parts per trillion) depending on instrumental sensitivity and frequency of occurrence in blanks. An *in vitro* cellular bioassay was used to determine the cumulative estrogenicity of compounds extracted from water samples, expressed as estradiol equivalents (EEq). The MRL of the bioassay was 0.16 ng/L, based upon the sensitivity of the assay, the volume of water extracted, and from the laboratory and field blank responses. Water samples were collected from 20 geographically diverse sites within the U.S. In total, more than 300 samples were evaluated during this investigation. Due to the high level of rigor needed for the risk assessment, more than 100 blanks and spiked water samples were analyzed in the course of this study.

To provide context for the EEq values determined in water samples, a series of food/beverage items were also evaluated. These items included vegetable juice, tea, coffee, beer, bottled water, soy sauce, milk, and baby formula. Food/beverage products were also analyzed for the suite of 62 target analytes. The food/beverage items were included in this study only to provide a reference point for relative source contributions. This was not a formal statistically-based market basket study and only a few of each type of product were evaluated.

Risk evaluations assuming exposure through drinking water were conducted for 16 pharmaceuticals, 10 potential EDCs, and three steroid hormones. Acceptable daily intakes (ADIs; also called "screening values") or reference doses (RfDs) were calculated using methods consistent with EPA approaches for determining levels of exposure to environmental contaminants that are not likely to be associated with adverse health effects. A cautious, conservative approach was taken in the process of developing the ADI values. These levels are commonly defined as the amount of a chemical a person (including sensitive subgroups or subpopulations) can be exposed to on a daily basis over an extended period of time (usually a lifetime) without suffering a harmful effect. Sensitive subgroups include those with chronic diseases, the elderly, children, pregnant women, and others.

For effects other than carcinogenicity, ADIs were derived by dividing the highest dose at which an effect was not observed (the "no observed adverse effect level" or NOAEL) or the lowest dose at which an effect was observed (the "lowest observed adverse effect level" or LOAEL) in animal or human toxicity studies by uncertainty factors (UFs) to account for extrapolation to potentially sensitive populations. Toxicity data were gathered through bibliographic searches, from information submitted to the U.S. Food and Drug Administration (FDA) as part of the drug approval process, or from drug labels or monographs. For analytes showing positive evidence of carcinogenicity in high dose animal studies and data on tumor incidence per dose level, a linear extrapolation model was used to predict the tumorigenic response at low doses. For analytes with reported evidence of carcinogenicity in animals but no available tumor incidence data, a "virtually safe dose" corresponding to a cancer risk of one in a million was estimated.

For some target EDCs, RfDs or ADIs have been previously established by regulatory agencies such as EPA and some have existing water quality criteria such as maximum contaminant levels (MCLs). However, most of these criteria were established before testing for endocrine mediated effects became standardized, and are based on non-endocrine toxicity endpoints. For the target EDCs, endocrine-mediated ADIs were established by identifying NOAELs and LOAELs for effects mediated through estrogenic, androgenic, or thyroid-related modes of action (MOAs). Emphasis was placed on data published since the establishment of the existing RfDs or ADIs. ADIs that account for endocrine mediated effects were estimated using methods similar to those described for non-carcinogens.

ADIs were converted to drinking water equivalent levels (DWELs) in units of micrograms per liter (µg/L, or parts per billion), by multiplying the ADI by an assumed body weight (70 kilograms, the EPA default adult body weight) and dividing by an average daily drinking water ingestion rate (two liters per day). For carcinogens, DWELs assume that a *de minimis* lifetime excess cancer risk of one additional cancer per one million lifetime exposures (10^{-6}) is acceptable.

RESULTS

Of the 62 target compounds investigated, only three were consistently (>50% frequency) detected in U.S. drinking water despite ng/L (parts per trillion) reporting limits. The herbicide atrazine exhibited the highest frequency of occurrence (83%) followed by the anti-anxiety pharmaceutical meprobamate (78%) and the anti-epileptic pharmaceutical phenytoin (56%). Atrazine also had the highest concentration detected at 870 ng/L, followed by the flame retardants TCPP and TCEP at 510 and 470 ng/L, respectively. It is important to note that the maximum detection of atrazine was less than 1/3rd of the federal drinking water regulation of 3,000 ng/L. Only 11 of the 62 target compounds were detected in >20% of the in finished drinking water. In order to gain perspective, consider that of these 11 compounds, only nine would have been detected using an MRL of 10 ng/L and only three would have been detected using an MRL of 100 ng/L. If the MRLs would have been one part per billion (1,000 ng/L), none of the target analytes would have been detected in this study. Again, the point is raised that what is detected is a direct function of how low the analytical method can measure. Therefore, as analytical procedures become progressively more sensitive, water utilities should expect that more pharmaceuticals and potential EDCs will be detected. Advanced treatment processes may provide initial results in lowering concentrations to non-detectable levels; however, as analytical methods are developed with lower MRLs, non-detectable compounds will likely again be detected. Even with current methodology and reporting limits, none of the pharmaceuticals tested in this study were detected in finished drinking water above the calculated health risk thresholds.

The maximum-detected concentrations in finished and distribution drinking water were used to calculate drinking water equivalent levels (DWELs) for each of the target pharmaceuticals. The maximum-detected concentrations used are the single highest discrete sample concentrations observed in this study, providing a conservative "worst-case" scenario approach. Using this approach, none of the pharmaceuticals detected in drinking water exceeded their corresponding ADI. Nine of the pharmaceutical target compounds were detected in finished drinking water. Two pharmaceutical compounds (norfluoxetine and risperidone) were also detected in distribution drinking water. The remaining nine pharmaceutical target compounds were not detected in either finished or distribution drinking water at the method reporting limits. Four compounds that were detected in either finished or distribution water (carbamazepine, gemfibrozil, phenytoin and risperidone) show some evidence of carcinogenicity in animal studies, and calculated ADIs and DWELs are based on cancer endpoints. Of the pharmaceuticals that were detected, three compounds (the two antibiotics/antibacterials, sulfamethoxazole and triclosan, and the anti-anxiety agent meprobamate) have DWELs greater than 200 μg/L and therefore appear to be of relatively low toxicity. Minimum margins of safety, (MOS) or margin of exposure (MOE), defined as the DWEL divided by the maximum-detected water concentration, were 6,000 for meprobamate, 6,000,000 for sulfamethoxazole and 2,200,000 for triclosan.. The MOS of three other detected compounds, diazepam (110,000), fluoxetine (41,000) and norfluoxetine (44,000), are also very large. The estimated potential hazard associated with exposure to the beta-blocker atenolol is also relatively low, with a DWEL of 70 μg/L and a minimum MOS of 2,700. According to US.EPA policy, and MOE greater than 100 for developmental effects would generally indicate a low level of concern., whereas a value less than 100 is judged to be of concern.

Using the same approach, the maximum detected drinking water concentrations of potential EDCs were compared to calculated DWELs. Only four of the 13 target EDCs were detected in finished drinking water at or above the method reporting limits. Using the worst-case approach described previously for determining the ADI and based upon the maximum drinking water concentration, none of the potential EDCs examined exceeded the calculated health risk threshold (i.e., DWEL). Atrazine has the lowest calculated DWEL, at 3 μg/L. The maximum detected concentration in finished drinking water was 1 μg/L, yielding a minimum margin of safety of three. Two compounds (bisphenol A and p-nonylphenol) have calculated endocrine-based DWELs greater than 200 μg/L and therefore appear to be of relatively low concern for endocrine-type toxicity based solely upon drinking water exposure. Minimum margins of safety for these compounds were 72,000 for bisphenol A and 16,000 for p-nonylphenol. The toxicity evaluation also indicates that the herbicide linuron is of relatively low concern for endocrine-mediated endpoints, with a calculated DWEL of 70 μg/L and minimum margin of safety of 8,400.

Estrogenicity as determined by the *in vitro* bioassay (E-screen) revealed that the vast majority of finished drinking waters were below the MRL. Only three finished drinking water samples from the entire study had estrogenicity greater than the reporting limit of 0.16 ng/L, with EEq values at 0.19, 0.20, and 0.77 ng/L. In comparison, several food and beverage items had far greater estrogenicity. For instance, vegetable juice had observed EEq values from 1.9 to 3.3 ng/L, while coffee ranged from 11 to 17 ng/L. Various brands of beer exhibited a broad range of results with EEq values ranging from 0.8 to 140 ng/L. The highest estrogenicity was observed in soy-based food and beverage items such as soy sauce (28 – 510 ng/L), soy baby formula (1,500 – 1,900 ng/L) and soy milk (1,900 – 4,200 ng/L). Results from estrogenicity testing of food products are provided simply as a comparative of the bioassay response and not as an indication of health risk. If a bioassay is to be used as a barometer for endocrine disrupting potential of drinking water, it is vital to understand how the assay responds to food and beverages we assume to be healthful.

RECOMMENDATIONS

Without question, the detection of pharmaceuticals and hormones in drinking water, even at extremely low concentrations, is likely to cause concern about the safety of water supplies. A number of factors could fuel this concern, such as the perception that people are being medicated unknowingly, the fact that many of the compounds originate from human waste, and the fact that some of these chemicals could cause health effects including cancer and developmental effects. Because of this, it is critical that the levels of exposure be placed into proper context. The reality is that nearly any chemical known to man could be detected in water using the most modern and sensitive of analytical instrumentation. All water on earth is reused over the course of time. Pharmaceuticals, hormones, and other chemicals in trace quantities are, and will continue to be, detectable in water. To make wise use of our economic and natural resources, these emerging contaminants of concern must be considered through the same mechanisms that are used to regulate other water contaminants.

The analysis of ultra-trace organic contaminants in drinking water is costly and may not lead to significant benefits in protecting public health. Even the most advanced and costly water treatment processes will not be able to removal all contaminants to less than the detection limit of the most sensitive analytical procedures. Therefore, it is imperative that toxicological relevance be established and considered to place concentrations of the various compounds into the context of human health. For utilities that wish to know whether pharmaceuticals and potential EDCs are present in their water sources, it is recommended that they choose a limited set of indicator chemicals as opposed to large suites of compounds. From the data presented in this study, and previous AwwaRF studies (e.g., AwwaRF project 2758), it is suggested that appropriate pharmaceutical indicators include meprobamate, phenytoin, atenolol, and carbamazepine. For steroid hormones, it is recommended that estrone and progesterone are the most useful indicators as they have the highest occurrence levels in a number of studies. In terms of potential EDCs, it is difficult to access the most reliable indicators since no formal studies regarding which of these chemicals are capable of exerting endocrine impacts to humans at levels relevant to drinking water have been conducted. The U.S. EPA's Endocrine Disruptor Screening Program (EDSP) is underway and will determine which chemicals are of greatest risk for endocrine disruption to the environment and to human health. It is recommended that utilities use *in vitro* bioassays carefully. While the bioassays are capable of extreme sensitivity, the link between cellular bioassay response and impacts to human endocrine function has not been established. Therefore, the detection of *in vitro* estrogenicity does not imply human health risk, nor does the absence of estrogenicity promise safety. This study shows the fallibility of using *in vitro* estrogenicity alone in assessing endocrine distupting potential, as some common dietary items contain far greater estrogenicity than impaired source waters. Certainly these food items are not considered a health risk. Considering that food items are not labeled, or often even tested, for emerging contaminants, it is difficult to argue that the choice of exposure from food is any less involuntary than would be exposure from tap water.

The evaluation of toxicological relevance provided here indicates that, although some pharmaceuticals and potential EDCs were detected in U.S. drinking waters, there is no evidence of human health risk from consumption of these waters. Furthermore, for the pharmaceuticals and potential EDCs detected in water, exposure to people through water is expected to be small compared to exposures to potentially hazardous compounds through prescription and non-prescription medications, food and beverages, occupational exposures, and residential activities (e.g., cleaning products, personal care products, hobby chemicals, pesticides). Moreover, the concentrations of some potential EDCs (e.g., plasticizers) are orders of magnitude greater in food products than in drinking waters, which may negate any appreciable gains achieved by advanced water treatment in attempts to mimization of human exposure.

It is recommended that utilities establish open communications within their organizations and to the public. Given the sensitivity of analytic detection instrumentation, utilities must try to help the public and key stakeholders understand that the presence of a compound in trace concentrations does not necessarily represent a health risk. Conveying risks related to this class of contaminants is conceptually no different than communicating with the public about the presence of other contaminants detected in water supplies. Information such as ADI or DWEL is critical to helping utilities make the necessary comparisons to put this issue in context. Utilities in the U.S. should stay abreast of developments within the federal and state regulatory agencies.

The study presented here is certainly not all inclusive or entirely comprehensive; however, it is intended to catalyze discussions and collaboration on additional needed studies among the water industry, regulatory agencies, and the public.

From this study, we recommend utilities consider the following points in emerging contaminant programs:

1. If monitoring is desired, consider a limited subset of indicator compounds that occur at higher concentrations and frequencies.

2. For monitoring programs, increase the frequency of blanks, replicates, spikes, and other QA/QC measures due to higher variability in ultra-trace methods.

3. Be sure that a part of any emerging contaminant program focuses on understanding what laboratory data mean from a human health standpoint.

4. Consider a strategic plan to understand which chemicals are of greatest current concern based on occurrence and potential health impact, and which ones may be on the horizon.

5. Be prepared to communicate the information obtained from monitoring studies and to help the public understand what the data mean.

6. Remain abreast of developments within local and federal regulatory agencies and participate as stakeholders whenever possible.

7. Follow the research in this area of emerging concern, and communicate key findings as they are developed. This should include continued addressing of emerging contaminants in appropriate meetings and conferences.

CHAPTER 1
INTRODUCTION

INTRODUCTION

All water on Earth contains measurable levels of various anthropogenic chemicals. The number and level of contaminants depends upon the factors influencing the water, the analytical methods applied, and the intensity of monitoring programs. In the past decade, a great deal of interest and concern has been generated regarding trace steroid hormones and pharmaceuticals in water (Daughton and Ternes 1999, Snyder et al. 1999b, Sedlak, Gray, and Pinkston 2000, Snyder et al. 2001c, Drewes, Heberer, and Reddersen 2002, Kolpin et al. 2002, Ternes et al. 2002, Metcalfe et al. 2003b, Westerhoff et al. 2005, Richardson 2007, Snyder et al. 2007a). While this era has seen a flurry of activity related to these emerging contaminants, the earliest published manuscripts regarding steroid hormones and pharmaceuticals in North American waters date back to the 1960s and 1970s (Stumm-Zollinger and Fair 1965; Tabak and Bunch 1970; Garrison et al. 1975; Hignite and Azarnoff 1977). Figure 1.1 demonstrates how research on steroids, pharmaceuticals and potential EDCs has increased over the past few decades. These compounds have increasingly become a concern for water quality control in drinking water supplies, either because of true toxicological implications or because of public perception.

One major contributor of these trace organic pollutants is wastewater discharge, which impacts surface water quality with chemical contaminants not completely removed by current wastewater treatment processes (Snyder et al. 2001a, Metcalfe et al. 2003a, Kolpin et al. 2004, Loraine and Pettigrove 2006, Voutsa et al. 2006). Indirect water reuse, where water influenced by wastewater discharge is used as source water for drinking water treatment plants (DWTPs), either planned or unplanned, is currently in practice in many areas. Indirect potable reuse will increase as shortages of water supplies become more prevalent and severe with the ever-increasing population. With such water reuse/recycling systems in place, potential for water contamination is greatly increased. Furthermore, with improvements in analytical sensitivity and selectivity, the terms “ubiquity” and “detection” begin to merge as chemists find traces of organic contaminants in a variety of aquatic environments, including seawater (Atkinson, Atkinson, and Tarrant 2003, Weigel et al. 2004), coastal sediments (Ferguson and Brownawell 2003, Braga et al. 2005), inland sediments (Mibu et al. 2004, Labadie and Hill 2007), and freshwater (Rodriguez-Mozaz, Alda, and Barceló 2004, Servos et al. 2007). The scientific community must grapple with the issue of the relationship between detection and risk. If a compound is detected in drinking water at parts per trillion concentrations (0.000000001 g/L), does that detection automatically imply a risk? As a corollary, if a compound is not detected in drinking water, is the population free of excess risk? With these issues in mind, this project was designed to examine the toxicological relevance of the presence of pharmaceutically active compounds (PhACs), endocrine disrupting compounds (EDCs), and personal care products (PCPs) in drinking water supplies.

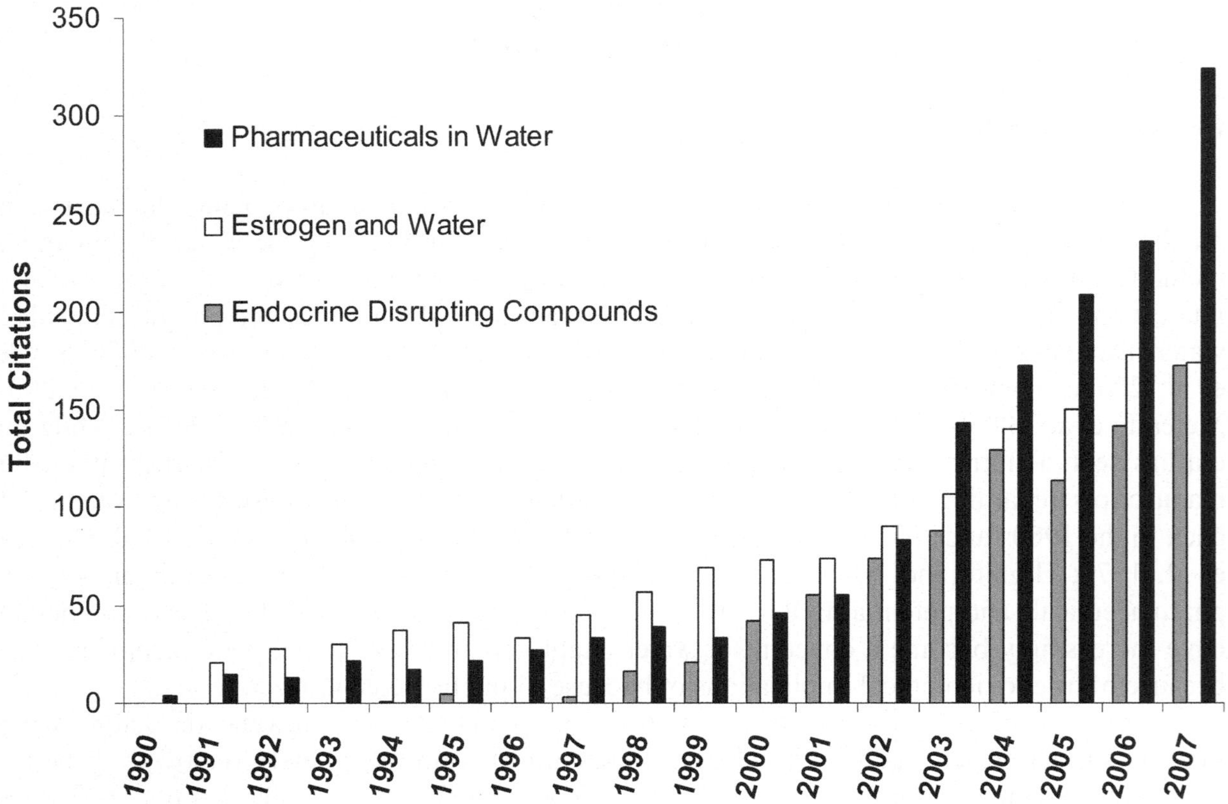

Figure 1.1 Number of citations by key words per year (1990 – 2007)

ENVIRONMENTAL OCCURRENCE

Emerging organic contaminants have been measured in wastewater effluent and receiving surface waters at levels sufficient to affect aquatic wildlife. In 1996, the first known North American report of endocrine disruption in fish below wastewater outfalls was published (Bevans et al. 1996). Since this time, there has been several other reports also documenting endocrine type effects on fish and other aquatic species exposed to wastewater effluents (Hemming et al. 2001a, Sole et al. 2003, Snyder et al. 2004, Gagne et al. 2007a). Tixier et al. 2003 studied three month variation in concentration of six pharmaceuticals in wastewater effluent and river waters (Tixier et al. 2003). The concentration profiles of these pharmaceuticals in the water column of a lake in Switzerland were measured over a time period of three months. Phototransformation, adsorption, and biodegradation were identified as the main elimination processes for these contaminants in the lake water. However, elimination of free and available species of contaminants in the water column is not an indemnity from effects upon wildlife species. For example, nonylphenols, an estrogenic biodegradation product of nonionic surfactants, have been shown to alter microbial populations in lake sediments (Jontofsohn et al. 2002). Many studies have identified impacts on wildlife species from short- and long-term exposure to sewage effluents and/or compounds present in sewage effluents (Hemming et al. 2001b, Aerni et al. 2004, Brion et al. 2004, Balch and Metcalfe 2006, Brian et al. 2007, Gagne et al. 2007b). Whereas wildlife may be directly exposed to water or contaminated soils and sediments at or near sewage outfall, direct contact of humans with surface water is likely to be minimal, with the exception of potential infrequent dermal contact or incidental ingestion during swimming or other recreational activities.

Even with multiple stages of treatment employed in drinking water treatment processes, PhACs and EDCs are still being detected in finished drinking waters (Servos et al. 2007a, Snyder et al. 2007b). Loraine et al. reported the occurrence of pharmaceuticals and personal care products in raw drinking water at four DWTPs that utilize water impacted by wastewater treatment plant (WWTP) effluent (Loraine and Pettigrove 2006). Several of the measured compounds were also detected in the finished drinking water. The concentrations in one of its source water showed seasonal variations during low and high flow conditions. Seasonal variation of five pharmaceuticals in Finland due to temperature change was reported in a drinking water plant which was impacted by WWTP effluent (Vieno, Tuhkanen, and Kronberg 2005). In another study, a DWTP with two-stage coagulation, followed by GAC filtration and chlorine disinfection confirmed previous findings obtained in laboratory research (Vieno, Tuhkanen, and Kronberg 2005, Vieno, Tuhkanen, and Kronberg 2006) that pharmaceuticals were poorly removed by coagulation, while GAC was very efficient in removing these contaminants (Snyder et al. 2007a; Adams 2002; Ternes et al. 2002; Vieno, Tuhkanen et al. 2006). Another study selected a traditional full-scale treatment plant, consisting of coagulation, filtration, and disinfection for the treatment efficiency of organic contaminants (Stackelberg et al. 2004). Pharmaceuticals and other organic wastewater-related contaminants were monitored during low flow conditions. Of the monitored compounds, 40 out of 106 were detected in raw or source water while as many as 17 were detected in finished water, indicating both the persistence of the compounds and the inefficiency of the treatment technology to remove the compounds. In perhaps one of the most comprehensive drinking water studies published to date, Snyder et al. 2007 (Snyder et al. 2007a) found traces of at least one contaminant (out of 31 compounds tested) in the finished water of each of 21 drinking water utilities tested in the U.S.

ANALYTICAL METHODS

Methods History

The advancement of analytical technology has led to increased detection of contaminants in the environment. As analytical methods and instrumentation improve, the levels at which compounds can be detected continue to decrease leading to more frequent reports of contaminant occurrence. Compounds such as estrogens, for example, were studied as early as the 1940s (Schueler 1946, Sluczewski and Roth 1948). In 1963, concentrations of 17β-estradiol were reported at low part per million (ppm) levels (Velle 1963). By the 1980s, estrogens were being reported at low part per billion (ppb) concentrations (Andreolini et al. 1987), tens to hundreds of part per trillion (ppt) in the 1990s (Shore, Gurevitz, and Shemesh 1993), and even lower into the low or sub ppt around the new millennium (Snyder et al. 1999a, Huang and Sedlak 2001a, Snyder et al. 2001b). With improved methodology, reporting limits of sub-ppt can readily be achieved with less sample volume and preparation than historic methods with ppm sensitivity. The newest generation of analytical technology provides the capability to simultaneously monitor stable isotopes of target analytes simultaneously with the actual contaminant (Trenholm et al. 2006; Vanderford and Snyder 2006). The future of contaminant monitoring will almost certainly rely on low sample volumes and automated extraction with direct interfaces to tandem mass spectrometric instruments to provide relatively high throughput using less solvents and significantly less preparation/analysis time (Barcelo, Postigo, and Alda 2008).

Compound Selection

The vast majority of analytical methods are based upon targeted contaminants that are known, or suspected to, occur in water. In the case of pharmaceuticals and potential EDCs, the number of potential compounds for evaluation is seemingly endless. Therefore, compound selection is critical and should be based upon three primary selection criteria. First, and most important, is protection of human health. In order to develop sensible emerging contaminant monitoring programs, compounds of greatest human toxicological concern should be selected as indicators. Corresponding method report limits should be at least as low as those levels determined to have no adverse human health impacts. Conversely, monitoring for trace contaminants at levels far below any reasonable human health protection, even after the application of conservative uncertainty factors (including additional factors for potential additivity and synergism) is wasteful of valuable resources and may lead to undue concern among the public. The U.S. EPA has developed a draft Candidate Contaminant List 3 (CCL3), which determines which unregulated drinking water contaminants are of priority for further evaluation for regulatory action. As of the writing of this report, the U.S. EPA has evaluated more than 1200 pharmaceuticals and added 287 to the Chemical Universe in the CCL3 draft (Federal Register February 21st, 2008). Of the 287 pharmaceuticals in the Chemical Universe, approximately 10% of these met the criteria to enter the preliminary-CCL (PCCL). Nitroglycerin is the only pharmaceutical that met the criteria to be listed on the draft CCL3; however, it is also noteworthy that nitroglycerin has far more applications beyond simply a human pharmaceutical. The final CCL3 will not be available until later in 2008 and certainly may change from the draft available at the writing of this report.

The selection of potential EDCs is far more daunting than the selection of pharmaceuticals. For the vast number of potential EDCs, very little (if any) human data are available. Therefore, guidance is quite sparse and limited mostly to speculation and opinions. However, certain chemicals have been widely labeled by the public, academic, and regulatory communities of potential EDCs of concern and are addressed in this report.

Phytoestrogens

Phytoestrogens, which are naturally occurring compounds present in plants (such as soybeans) or plant products (such as fruits), can act as estrogens in the body. Naturally occurring phytoestrogens drew attention several decades ago when a positive relation between cow mastitis and phytoestrogens was discovered (Brookbanks, Welch, and Coup 1969a). The majority of phytoestrogens are from the flavonoid group, which is subdivided into flavone, flavonol, flavanone, and flavanol types according to the presence and position of carbonyl and hydroxyl groups at C-3 and C-4 and a saturated single bond between C-2 and C-3 (Prasain, Wang, and Barnes 2004) (see skeleton in Figure 1.2). Isoflavonoids are flavonoids with ring B attached to the C-3 position of ring C, instead of C-2 position. Genistein and daidzein are two of the most abundant and most studied isoflavonoids that have shown some health benefits due to their anticarcinogenic, antiatherogenic and antioxidative potentials. The other two main groups are lignans (found in flaxseed) and coumestans (derived from sprouting plants like alfalfa). Legumes (such as soybean) constitute a significant source for human phytoestrogen dietary intake of both genistein and daidzein; while orange juice, milk and sometimes beer are an important source for biochanin A and formononetin. Among the list, equol is a metabolite from legumes and commonly detected at relatively high concentrations. Because of the significant differences in prostate and breast cancer incidence between western and eastern diets (Dhom 1992), it has long been suspected that phytoestrogens, though very weak estrogens, help prevent the disease. A great amount of research has been conducted to study phytoestrogens, which in general fall into two areas: occurrence and health effects.

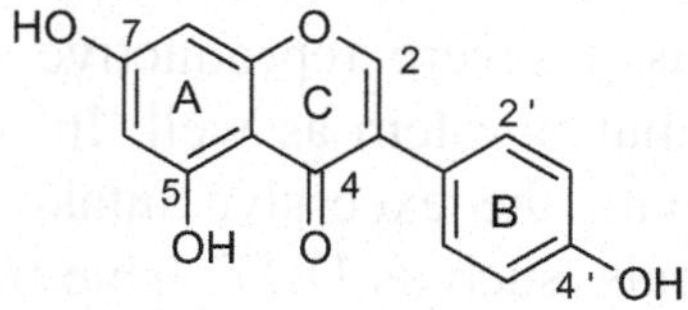

Figure 1.2 Structure of Genistein

Phytoestrogens have been shown to have both beneficial and adverse health effects in animals depending on the dose. The low incidence of breast cancer in Asian women has been explained by their high intake of phytoestrogens (Adlercreutz et al. 1991). Case-control studies indicate the protective effect of isoflavones intake on breast cancer (Wu et al. 1996, Ingram et al. 1997, Murkies et al. 2000, Lu et al. 2001, Shu et al. 2001), which can even be traced back to phytoestrogen intake during adolescence (Shu et al. 2001). The protective effect from phytoestrogens is supported by overwhelming evidence collected from both *in vivo* and *in vitro* data (Tsuda et al. 2003). Possible anticarcinogenic effects of phytoestrogens in breast cancer were attributed to their ability to bind to estrogen receptor α and β and exert antiestrogenic effects through inhibitory binding (Tang and Adams 1980, Waters and Knowler 1982, Adlercreutz 1984, Setchell et al. 1984, Kuiper et al. 1998). While a preponderance of data show the health benefits of phytoestrogens, studies also have shown that estrogenic effects can be significant in high doses. In humans, phytoestrogens in beer have been associated with adverse health effect in men who consume large volumes of alcohol. Depending on the time after drinking and the amount of alcohol consumed, higher level of estradiol and lower level of dihydrotestosterone was observed, while testosterone levels were either depressed or elevated (Couwenbergs 1988). Recently, a novel phytoestrogen, 8-prenylnaringenin, was discovered in beer that is significantly more potent than those previously reported phytoestrogens (Pepper et al. 2004). Phytoestrogens are commonly detected in infant milk formula and even cow's milk. Their concentrations are comparable to estrogen steroids and sometimes even higher. For instance, the detected steroid concentrations range from 20 to 116 ng/mL in human milk (Sahlberg and Axelson 1986) and up to 8.6 ng/L (17β-estradiol) in cow's milk (Lopez, Satter, and Wiltbank 2004). Phytoestrogen concentrations in powdered milk range from 7 to 475 ng/g, in soybean milk from 2.5 to 212,000 ng/g and in cow's milk from 1 to 30ng/mL (Antignac et al. 2003). Soy based infant formula may constitute a significant source of phytoestrogens for babies, especially neonates. An average of 42 ppm phytoestrogens was observed in five soy based infant formulas. In fact, phytoestrogens start playing a role during pregnancy since they are frequently detected in the amniotic fluid of pregnant women (Foster et al. 2002). The plasma concentrations of genistein and daidzein are 100 to 300 times higher in infants fed with soy milk formula than in those fed with cow's milk formula or breast milk (Setchell et al. 1997). These levels are also an order of magnitude higher than the levels known to have hormonal effects in adults and 13,000 to 22,000 times higher than plasma estradiol concentration in early life (Setchell et al. 1997). Phytoestrogens delivered via soy milk to rats from birth until weaning have caused significantly higher concentrations of progesterone receptor in 2-month-old pups despite insignificant effects to reproductive organs (Hughes et al. 2004). It has long been shown that phytoestrogen rich diet results in infertility and mastitis of sheep (Brookbanks, Welch, and Coup 1969a, Adams 1998). A syndrome known as clover disease in sheep was related to the estrogenic activity of isoflavones (Bennetts, Underwood, and Shier 1946, Moule, Braden, and Lamond 1963, Braden, Hart, and Lamberton 1967, Shutt and Braden 1968), which originates in pastures containing red clover. The syndrome is shown as a severe reproductive disorder that results in permanent infertility. Other animals have similar problem as well. In one study, the decreased fertility in captive cheetahs was associated with the excessive intake of phytoestrgoens (Setchell et al. 1987). Synthetic estrogenic chemicals, such as DDT, have been blamed for low sperm count in human males, but Safe pointed out that the effect of natural estrogens is 40,000,000 times higher than estrogenic pesticides when daily intakes were considered (Safe 1995).

PHARMACEUTICALS AND RISK IN DRINKING WATER

One of the major concerns over long term low dose exposure of pharmaceuticals is the increased resistance of bacteria to antimicrobial products and allergic reactions in humans associated with exposure to trace levels of PhACs, EDCs, and PCPs (Schwarz, Kehrenberg, and Walsh 2001, Phillips et al. 2004, Turnidge 2004). Endocrine disrupting chemicals and hormone steroids are known to interrupt endocrine system functioning. Phytoestrogens are naturally occurring compounds present in plants, such as soybeans and plant products, which act like estrogens in the human body. Naturally occurring phytoestrogens have drawn attention for several decades when a positive relation between cow mastitis and phytoestrogens: formononetin-like and genistein-like substances was discovered (Brookbanks, Welch, and Coup 1969a). Drugs are designed to be biologically active. While beneficial for one population group, the same biological activity can be hazardous to other population groups or at certain times—for example, during pregnancy or while being exposed concurrently to other compounds. Additionally, little is known about the potential health impacts of long-term, low level exposure to pharmaceuticals, such as may occur if pharmaceuticals are present at trace levels in drinking water.

Several investigators have examined the potential human health impacts of subsets of pharmaceuticals. Based on information about the activity and potential toxicity of pharmaceuticals, it has been suggested that possible health effects of long-term exposure to pharmaceuticals in drinking water could include endocrine disrupting activity, induction of antibiotic resistance, genotoxicity, carcinogenicity, or allergic reactions, as well as effects on reproduction or fetal/ child development (Daughton and Ternes 1999). Other researchers assessed the potential for adverse health effects from exposure to pharmaceuticals in drinking water by comparing exposures from drinking the water to the therapeutic dose, divided by uncertainty or safety factors to extrapolate from these doses to safe levels for populations that could include sensitive individuals (Sabaliunas, Webb et al. 2001; Webb, Ternes et al. 2003; Schwab, Hayes et al. 2005). However, while therapeutic doses can be useful benchmarks, uncertainty exists about the potential effects of these pharmaceuticals on nontarget population groups (such as the fetus, infants, or the elderly), or about the effect of such factors as differences in dosing and timing of exposure or exposure to multiple chemicals at the same time (Jones, Voulvoulis, and Lester 2004).

Other authors examined human health risks of a limited number of pharmaceuticals (antibiotics, antineoplastics, and non-steroidal anti-inflammatory drugs (NSAIDs)) in drinking water based on no effect levels from animal or human toxicity tests; they concluded that no appreciable risk for humans exists at the low levels measured in drinking water (Christensen 1998, Schulman et al. 2002). Recently calculated provisional "no-effect levels" for a number of pharmaceuticals detected in Dutch drinking water, based on acceptable daily intake (ADI) or maximum residue levels (MRL) for veterinary pharmaceuticals in milk (Versteegh et al. 2003). These investigators drew similar conclusions, namely that detected concentrations were far below the levels of concern for human health.

Although increasing data are available on PPCPs and EDCs in water systems, testing of water supplies for PPCPs and EDCs is not yet systematic, and available measurements are variable and widely distributed. Most sampling programs to date have not been designed to specifically assess concentrations at locations where people are known to come in contact with the water. In general, data that are collected provide qualitative information on whether PPCPs or EDCs are present, and to a lesser extent, quantitative information on actual human exposure through drinking water.

REGULATIONS

Although there are no formal Federal regulations in the U.S. limiting the concentrations of pharmaceuticals and most potential EDCs, the 1996 Amendments to the Safe Drinking Water Act explicitly dictate that the U.S. EPA must "develop a screening program, using appropriate validated test systems and other scientifically relevant information, to determine whether certain substances may have an effect in humans that is similar to an effect produced by a naturally occurring estrogen, or other such endocrine effect as the Administrator may designate." The EPA's Endocrine Disruptor Screening Program (EDSP) was developed to meet the Safe Drinking Water Act requirement and is underway. The EDSP provides a comprehensive evaluation of the endocrine disruptive effects of chemicals which have the propensity to contaminate food and drinking water. Through a tiered battery of in vitro and in vivo bioassays, the endocrine disrupting potential to both the environment and human health will be evaluated. The validation of appropriate assays to assess human and environmental impacts of EDCs was an extremely arduous task. The belated schedule of the EDSP was partially related to the critical need for reproducibility and transferability. Without a rigorously validated battery of assays, the program would be of little value and would be less protective of environmental and human health.

As discussed previously, the U.S. EPA has a mechanism for the evaluation of unregulated contaminants for potential regulatory consideration. The EPA should be commended for the diligent work on this topic despite resource limitations for evaluation of the "universe" of potential drinking water contaminants. While federal regulations specifically designed for pharmaceuticals and potential EDCs in drinking water are essentially non-existent, the CCL process provides an evaluation of unregulated contaminants including those with exceptional notoriety. Beyond federal regulations, individual states may under primacy agreements create more stringent or even additional regulations. For instance, California and Massachusetts have established enforceable maximum contaminant levels (MCLs) for perchlorate in drinking water based upon disruption of the thyroid gland. Thyroid is one of three primary endocrine endpoints as defined by the U.S. EPA's EDSP. Therefore, utilities should be mindful that despite limited regulatory activity by the Federal government in relationship to pharmaceuticals and EDCs, it is plausible that individual states will continue to develop enforceable regulations for contaminants of emerging concern not regulated on a national level.

CHAPTER 2
SELECTION OF TARGET ANALYTES

SELECTION OF TARGET ANALYTES – PHARMACEUTICALS

Criteria

More than 3,000 pharmaceuticals are currently approved for prescription in the U.S. Additionally, hundreds of other compounds are approved for over the counter (OTC) use, constituents of personal care products and fragrances, or used as adjuncts in the formulation of these materials. Because of resource limitations, only a small subset of indicator pharmaceuticals were included for evaluation in this investigation. Several selection criteria were identified to help focus the compound list on relevant compounds. These are as follows:

Criteria 1: Prescription Pharmaceuticals. Only prescription, as opposed to OTC pharmaceuticals were selected because they tend to have greater toxicity potential, particularly to sensitive population groups. For example, per (FDA 2006), a pharmaceutical *can not* be classified for OTC use if it has an unacceptable margin of safety (*i.e.*, they are potentially toxic at therapeutic doses) or a significant potential for harmful effects or misuse/abuse. Pharmaceuticals that primarily affect the hormone system (*e.g.*, contraceptives, corticosteroids, thyroid agents, estrogen replacements, diabetes treatments) were considered for evaluation as potential EDCs.

Criteria 2: Volume of Use. The Top 200 most prescribed pharmaceuticals in 2003 and the Top 300 most prescribed pharmaceuticals in 2004 (NDCH 2005) were considered. Pharmaceuticals fell within one of these groups if approximately 2.2 million or more prescriptions were administered per year in the U.S.

Criteria 3: Toxicity. Screening toxicity data were reviewed to identify those pharmaceuticals with the most potential to cause adverse health effects at very low, chronic exposure levels. In particular, information was sought on evidence of carcinogenicity, mutagenicity, impairment of fertility, developmental toxicity, immunotoxicity, and neurotoxicity in animal or human studies. Toxicity information was extracted primarily from summary toxicity data presented by the U.S. FDA in their Drugs@FDA online catalog of approved pharmaceutical products (www.accessdata.fda.gov/scripts/cder/drugsatfda/) or in the Physicians Desk Reference (NDCH 2005). Each pharmaceutical's U.S. FDA "Pregnancy Category" was also noted. U.S. FDA classifies all approved prescription pharmaceuticals according to pregnancy category as listed in Table 2.1.

Table 2.1
U.S. FDA Pregnancy Categories

Category	Definition	Clinical application
Category A	Controlled studies in women fail to demonstrate a risk to the fetus in the first trimester, no evidence of risk exists for later trimesters, and the possibility of fetal harm appears remote.	For all practical purposes, there are no Category A pharmaceuticals.
Category B	Either (a) Animal studies have not shown a risk to the fetus, but no adequate studies have been done in pregnant women, or (b) Animal studies have shown a risk to the fetus but human studies show no risk in the first trimester and no evidence exists of risk in later trimesters.	If there is a clinical need for a Category B pharmaceutical, it is considered safe to use during pregnancy.
Category C	Either (a) Animal studies have shown an adverse effect on the fetus, but no adequate studies have been done in humans although the benefits may outweigh the risks; or (b) No animal studies exist and any human studies are inadequate to judge risk.	If the pregnant patient will benefit from a Category C pharmaceutical, it is generally used, although Category B pharmaceuticals are preferred if they will give equivalently good results.
Category D	Data from humans provide some evidence of fetal risk, but the benefits of use may outweigh risks (*e.g.*, the pharmaceutical is needed in a life-threatening situation or for a serious disease).	Category D pharmaceuticals have some significant risks; they are to be used during pregnancy only when alternatives are worse.
Category X	Studies in animals or humans have shown fetal abnormalities and the risk of use in pregnant women clearly outweighs any possible benefit.	Category X pharmaceuticals should not be used during pregnancy.

Source: 21 CFR § 201.57

Candidates for further evaluation were selected based on screening toxicity information as follows:

- All pharmaceuticals in Pregnancy Category "D" or "X"
- All pharmaceuticals with evidence of carcinogenicity in animal studies
- All pharmaceuticals with a "Yes" in four or more toxicity categories (a "Yes" was assigned to a category if positive evidence of the pharmaceutical causing mutagenicity, carcinogenicity, impairment of fertility, developmental toxicity, immunotoxicity, or neurotoxicity in animals or humans was noted)

While limited information on metabolism and excretion was also compiled to assess whether a pharmaceutical is likely to be excreted from the body in a chemically active form, this information is not necessarily indicative of which pharmaceuticals might be detected in water systems because non-metabolized pharmaceuticals can find their way into water systems through other routes (*e.g.*, flushed down the toilet, released in manufacturing effluent) or because the unmetabolized fraction of the pharmaceutical, once released, is chemically stable and persists in the environment, such that concentrations tend to increase. For example, although less than 2% of gemfibrozil is reportedly excreted from the body unchanged (NDC 2005), gemfibrozil is detected frequently in water systems at concentrations among the highest of any pharmaceutical (*e.g.*, up to 5,500 ng/L in sewage treatment plant outflows in the U.S.; (Sedlak, Pinkston, and Huang 2005), Appendix B). Likewise, very little phenytoin (Dilantin) is excreted from the body unchanged (FDA 2000), but phenytoin has been detected at concentrations up to 261 ng/L in the Las Vegas Bay of Lake Mead, Nevada (Snyder et al. 2001d). Thus, though informative, data on metabolism and excretion was not used as a primary basis to exclude a pharmaceutical from further consideration.

Criteria 4: Occurrence and Public Interest. Data on detection frequency in drinking waters or sources, as cited in published journal articles or agency reports, also was considered. This information was used to help select between pharmaceuticals of similar apparent toxicity. However, because a relatively small number of pharmaceuticals have been monitored in water systems, lack of measurement data was not used as a primary reason for excluding a compound.

Conversely, several pharmaceuticals that did not pass the other screening criteria were selected for the detailed assessment because they have been detected with relatively high frequency in water systems and have attracted public and regulatory interest, and because they could serve as useful indicator chemicals for the presence ofother pharmaceuticals in water. Example compounds include carbamazepine, naproxen, sulfamethoxazole, and trimethoprim.

Criteria 5: Pharmaceutical Group. Since only a limited number of pharmaceuticals could reasonably be analyzed in this assessment, representative pharmaceuticals of specific pharmaceutical groups were selected (*e.g.*, angiotensin-converting enzyme (ACE) inhibitors, antilipidemics including fibrates and statins, anticonvulsants, antipsychotics, benzodiazepine tranquilizers, selective serotonin reuptake inhibitors). Within particular groups, an effort was made to include compounds of relatively higher potential toxicity, unless other factors (*e.g.*, analytical issues; see Criteria 6) precluded analysis in water samples gathered in this assessment.

Criteria 6: Analytical Capability. Some compounds that passed other screening criteria were ultimately excluded from the target compound list due to lack of adequate calibration standards to ensure the accuracy and precision of applied analytical methods. In some cases, the developed analytical methodologies captured a significant number of desired analytes, but was less successful for certain outlier compounds. In these cases, the decision to sacrifice a few for the value of many was necessary. These excluded compounds include ciprofloxacin, divalproex (Depakote®), lisinopril (Prinivil® or Zestril®), olanzapine (Zyprexa®), and sertraline (Zoloft®).

The non-EDC pharmaceuticals considered in this assessment are listed in Appendix C, with summary information on their usage rate, pharmaceutical group, occurrence, and toxicity, and the rationale for including or excluding the compound in the final target compound list. The base list is comprised of the Top 200 non-EDC pharmaceuticals from 2003, with the addition of the three non-EDC pharmaceuticals in the Top 200 in 2004 that were not on the 2003 list. When identical formulations were listed under different brand names, only the brand with the largest number of total prescriptions in a given year was included in the list.

The following compounds were selected for inclusion as analytes:

Atenolol (Tenormin®, prescribed largely as a generic), a beta-blocker used in the treatment of high blood pressure — Selected because of its potential toxicity to the unborn fetus and evidence of carcinogenicity, and as a representative of this widely prescribed class.

Atorvastatin and metabolites o-hydroxy atorvastatin and p-hydroxy atorvastatin (Lipitor®), an antilipidemic or "statin" used in the treatment of high cholesterol and the most-prescribed pharmaceutical in the U.S. in 2003, 2004, and 2005—Selected because of its potential toxicity particularly to the unborn fetus, potential carcinogenicity, and high level of use.

Carbamazepine (Tegretol®), an anticonvulsant used in the treatment of seizures, nerve pain, and bipolar disorder—Selected because of its potential toxicity particularly to the unborn fetus, and very frequent detection in water systems world wide.

Diazepam (Valium®), a benzodiazepine anti-anxiety agent—Selected as a representative of this highly prescribed class because of somewhat higher potential toxicity than other more-prescribed class members such as alprazolam (Xanax®) and lorazepam.

Diclofenac (prescribed largely as a generic), a non-steroidal anti-inflammatory pharmaceutical (NSAID) used in the treatment of pain and inflammation—although not a candidate based on Criteria #3 (toxicity screen), selected because of very frequent detection worldwide and to retain a representative of this widely used class.

Enalapril (Vasotec®), an ACE inhibitor used in the treatment of high blood pressure—Selected because of potential toxicity to the unborn fetus and occasional detection in U.S. waters, and as a representative of this frequently prescribed class (seven ACE inhibitors were in the Top 200 most prescribed pharmaceuticals in 2003). The most prescribed pharmaceutical of this class, lisinopril, was excluded because of problems acquiring an analytical standard.

Fluoxetine and metabolite norfluoxetine (Prozac®), a selective serotonin reuptake inhibitor (SSRI) antidepressant—Selected because of potential toxicity to the unborn fetus, wide use, and occasional detection in U.S. waters. The most prescribed pharmaceutical in this class, sertraline (Zoloft®), was excluded because of problems acquiring an analytical standard.

Gemfibrozil (Lopid®), an antilipidemic used to treat high cholesterol—Selected because of its potential toxicity to the unborn fetus, evidence of carcinogenicity, and frequent detection worldwide.

Naproxen (Aleve®, available by prescription and OTC), an NSAID used in the treatment of pain and inflammation—Selected because of frequent detection worldwide.

Phenytoin (Dilantin®), an anticonvulsant used in the treatment of seizure disorders and epilepsy, as well as mood disorders—Selected because of its potential toxicity to the unborn fetus and evidence of carcinogenicity.

Risperidone (Risperdal®), an atypical antipsychotic used in the treatment of schizophrenia, bipolar disorder, and Tourette syndrome—Selected because of its potential toxicity to the unborn fetus, evidence of carcinogenicity, and as a representative of this increasingly prescribed class. A more frequently prescribed member of this class, olanzapine (Zyprexa®), was excluded because of problems acquiring an analytical standard.

Simvastatin and metabolite simvastatin hydroxy acid (Zocor®), an antilipidemic or "statin" used in the treatment of high cholesterol—Selected because of its potential toxicity to the unborn fetus, evidence of carcinogenicity, and as a representative of this widely prescribed class. Simvastatin was among the top four most-prescribed pharmaceuticals in the U.S. in 2003, 2004, and 2005.

Sulfamethoxazole, a generic antibiotic typically used in formulations with trimethoprim— Although not a candidate based on Criteria #3 (toxicity screen), selected because of very frequent detection worldwide.

Trimethoprim, a generic antibiotic typically used in formulations with sulfamethoxazole— Although not a candidate based on Criteria #3 (toxicity screen), selected because of very frequent detection worldwide.

Two additional compounds not on the lists of most prescribed pharmaceuticals were included because of frequent detections, even though they were not candidates based on Criteria #3 (toxicity screen):

Meprobamate, an antianxiety agent in use since 1955— Selected because it is widely detected in water systems due in part to a high degree of resistance to oxidation and shows some evidence of risk to the developing fetus.

Triclosan, a broad spectrum antibacterial/antimicrobial agent frequently used in handsoaps— Selected because it is the subject of increasing concerns/debates regarding its potential to cause antibiotic resistance.

SELECTION OF TARGET ANALYTES – POTENTIAL EDCS

Hundreds of chemicals have been purported to be endocrine disruptors (EDCs) or more appropriately potential EDCs based on a variety of different criteria, ranging from hormone receptor binding activity to cellular responses in vitro to stronger evidence of endocrine activity in vivo. For example, the Institute for Environment and Health (IEH 2005) in the United Kingdom identified 966 chemicals or elements that have been suggested in the published literature to be potential endocrine disrupters. Known or putative EDCs include hormones excreted by humans and animals, phyotestrogens and mycoestrogens, naturally-occurring ions (e.g., perchlorate) and metals (often mobilized in the environment by human activities), human and veterinary pharmaceuticals, pesticides and other biocides, industrial chemicals and their degradation products, and combustion byproducts from natural (volcanic activity) and anthropogenic (vehicle exhaust, incinerators) sources.

Selection Criteria

Because it was not feasible to conduct a review of occurrence data or toxicity data for all potential EDCs for this project, a list of chemicals of potential concern that fall under the category of EDCs was selected using the criteria described below. Chemicals were selected using an iterative process such that thorough literature reviews were not completed for each of the criteria for each chemical when it became apparent that a chemical did not meet one of the criteria. For example, if it became clear during the process of evaluating a chemical that it was unlikely that a reliable analytical method could be developed or applied by Southern Nevada Water Authority's (SNWA) Applied Research & Development Center with a reasonable amount of effort and there were few or no occurrence data in the published literature, literature searches for toxicity data were halted and efforts redirected toward the remaining candidate chemicals.

Criterion 1: Status as an EDC.

Chemicals with the potential to interfere with endocrine systems are called endocrine disrupting chemicals (EDCs). Other terms used to describe these chemicals are "endocrine disruptors," "hormonally active agents", "endocrine active chemicals" or similar variations of these. Various agencies and organizations use different definitions or descriptions for this class of chemicals. Participants in a U.S. EPA-sponsored workshop defined EDCs "exogenous agents that interfere with the production, release, transport, metabolism, binding, action, or elimination of the natural hormones in the body responsible for the maintenance of homeostasis and the regulation of developmental processes" (Kavlock et al. 1996). The International Programme on Chemical Safety (IPCS) adopted the following definitions for endocrine disruptors: "An endocrine disruptor is an exogenous substance or mixture that alters function(s) of the endocrine system and consequently causes adverse health effects in an intact organism, or its progeny, or (sub)populations. A potential endocrine disruptor is an exogenous substance or mixture that possesses properties that might be expected to lead to endocrine disruption in an intact organism, or its progeny, or (sub)populations" (WHO-IPCS 2002a). IPCS requires not only an effect, but specifies an adverse effect, and the effect must occur in an intact organism rather than in vitro.

After much debate among its members, the Endocrine Disruptor Screening and Testing Advisory Committee (EDSTAC) to the U.S. EPA's Endocrine Disruptor Screening Program (EDSP) decided, "…an endocrine disruptor as an exogenous chemical substance or mixture that alters the structure or function(s) of the endocrine system and causes adverse effects at the level of the organism, its progeny, populations, or subpopulations of organisms, based on scientific principles, data, weight-of-evidence, and the precautionary principle" (EDSTAC 1998).

For a chemical to be classified as an EDC, the U.S. EPA (Timm 2007) and the European Commission (EC 2002) further stipulate that it's critical effect must be one that results from a primary assault on the endocrine system rather than secondary or other more conventional toxicologic effects. That is, the endocrine-mediated effect should occur at a dose smaller than that required to cause general systemic toxicity. Likewise, according to the GWRC (GWRC 2003), "An endocrine disrupter could be defined as 'a substance that causes perturbation of hormonal/endocrine status at dose levels which are not cytotoxic.' In other words, substances that cause changes in endocrine status only at doses that also cause tissue damage would not be considered."

The U.S. EPA developed the EDSP to identify screening methods and toxicity testing strategies that can be used to determine whether a chemical should or should not be considered to be an EDC acting through estrogenic, androgenic, or thyroid-related (EAT) modes of action. Many chemicals have not been tested by any method for potential endocrine activity, and controversy surrounds assigning the label "EDC" to many products of commercial value. However, certain chemicals are widely considered to be EDCs based on effects reported in humans, laboratory animals, and/or fish and wildlife (Table 2.2). Still, not all experts agree that each of these chemicals should be classified as an EDC.

Table 2.2
Selected examples of chemicals widely considered to be EDCs based on effects in laboratory animals and/or in fish and wildlife

Class/Description	Selected Example Chemicals
Hormones	▪ 17ß-Estradiol ▪ Estrone
Pharmaceuticals	▪ Ethynylestradiol ▪ Diethylstilbestrol (DES)
Industrial chemicals or their degradation products	▪ Bisphenol A ▪ Nonylphenol ▪ Octylphenol ▪ Perchlorate ▪ Certain phthalates: di-n-butyl phthalate, butylbenzyl phthalate, di-2-ethylhexyl phthalate, others ▪ Certain PCBs*
Biocides (pesticides, herbicides, fungicides, etc.)	▪ Atrazine ▪ DDT† & some of its degradation products ▪ Linuron ▪ Methoxychlor ▪ Vinclozolin

* PCBs = polychlorinated biphenyls
† DDT = dichlorodiphenyl trichloroethane

Many chemicals might be expected to cause endocrine disruption based on results of screening-level tests such as quantitative structure-activity relationships (QSARs) or effects in some in vitro testing systems. Although this preliminary evidence might indicate the potential for endocrine activity, the more reliable and relevant in vivo tests do not always allow the conclusion that a chemical is indeed an EDC. In comparison with the number of chemicals that have been screened for endocrine activity, relatively few chemicals have been subjected to the more definitive in vivo testing. Also, while a positive result for endocrine activity in an in vivo test provides strong evidence that a chemical is an EDC, a negative result does not preclude the possibility that a chemical might have endocrine activity in a different test system. While EDSP is currently developing and validating both screening assays and more conclusive bioassays that can be used to identify EDCs, no scientific consensus has yet been reached within the EDSP or other expert groups regarding the specific suite of tests that should be used to determine whether a contaminant is or is not an EDC. However, the Endocrine Disruptor Methods Validation Subcommittee (EDMVS, later renamed the EDMVAC—Endocrine Disruptor Methods Validation Advisory Committee) of the EDSP agreed that designation as an endocrine disruptor should require that a chemical is associated with an adverse effect in a Tier 2 study. Tier 2 studies include only in vivo tests conducted according to specified protocols and do not include all of the studies available in the open literature because of potential variability. U.S. EPA will accept Tier 2 assays as definitive for risk assessments.

The U.S. EPA is cooperating with the European Organisation for Economic Co-operation and Development (OECD), which also is developing methods to identify EDCs. However, standardization of tests and selection of a battery of tests are tasks that are not expected to be completed for several years. Initial screening of a list of pesticide ingredients is expected to be required under the EDSP in early 2008 (USEPA 2007b), but the more definitive Tier 2 tests are still under development. Recently, on October 16, 2007, the OECD adopted its Test Guideline for the Uterotropic Bioassay, and the organization anticipates adopting several other test guidelines for EDCs in 2008 (Musset 2007).

For the purposes of the current study, emphasis was placed on chemicals that have produced an adverse effect mediated through the endocrine system in at least one in vivo test system using a laboratory animal serving as a surrogate for humans (e.g., rats, mice, guinea pigs, rabbits, non-human primates) or that have been reported to produce an endocrine effect in humans, regardless of whether the effect is considered negative or therapeutic (e.g., ethynylestradiol, diethylstilbestrol, phytoestrogens).

Although standardized toxicity tests (particularly reproductive and developmental tests) sometimes can identify effects that occur by an endocrine mode of action (MOA), most currently available standardized in vivo tests do not include endpoints that allow an assessment of whether the effect occurs through endocrine disruption. Thus, in most cases, the results of a single test alone are not enough to conclude that an effect occurs through an endocrine MOA and that the endocrine-mediated effect is the critical effect. In those cases, a weight-of-evidence approach is recommended to evaluate whether a chemical exerts its critical effects as an EDC. A weight-of-evidence approach and examples of its use are provided elsewhere (WHO-IPCS 2002b).

The following reports were used to screen the initial list of potential EDCs to identify those for which there is stronger evidence of potential for in vivo effects mediated through the endocrine system:

Towards the Establishment of a Priority List of Substances for Further Evaluation of Their Role in Endocrine Disruption: Preparation of a Candidate List of Substances as a Basis for Priority Setting (EC-BKH 2000). This report describes the development of a priority list as follows. First, a working list of chemicals purported to be EDCs was compiled from lists of potential EDCs published by various organizations and then supplemented with a limited review of the scientific literature to identify reports and papers describing effects suggestive of endocrine disruption. A stakeholder meeting was held to ensure that the list was comprehensive. The information gathered in this process was used to identify those chemicals that might be either highly persistent in the environment or that are produced by industry in large volumes (high production volume, or HPV, chemicals). Finally, the strength of evidence for endocrine disruption and likelihood of exposure were assessed. For the purposes of the current project, interest was focused on chemicals labeled as Group I chemicals. These are chemicals for which there is at least one study that shows endocrine disruption in an intact organism (Category 1 designation) and which are further assigned to the high exposure concern group. High exposure concern means that "...human exposure is expected, due to environmental concentrations and those in food or consumer products, also taking into consideration exposure of vulnerable groups *And/Or* Wildlife exposure is expected, due to use and emission patterns, and the chemical is persistent and bioaccumulative."

Category 1 and Group I designations do not differentiate between potential for effects on humans and effects on other organisms. Appendix 13 of the EC-BKH (2000) document was examined to identify chemicals in Category 1 that were assigned a ranking of 1 for potential human health effects. The current project is concerned only with potential human health effects through drinking water, but the EC-BKH (2000) exposure concern designation is based on many potential routes of exposure. For this reason, additional information (other than that presented in Appendix 13) is needed to judge whether exposure via drinking water is likely. The priority list developed by EC-BKH (2000) is "based on a screening of available literature and should therefore be regarded as a starting point for further in-depth evaluation of the substances placed on the priority list...The evaluations... are NOT considered comprehensive risk assessments."

Chemicals Purported to be Endocrine Disruptors: a Compilation of Published Lists (IEH 2005). This report compiles and summarizes various reports and listings produced by national or international governments and non-governmental organizations as well as supplementary data taken from original studies or review articles identified in the peer-reviewed literature published from 2000 – 2002. These efforts resulted in a consolidated listing of 966 chemicals or elements that have been suggested in the published literature to be potential endocrine disrupters. This report identifies chemicals designated in EC-BKH (2000) Group I (see above).

Global Assessment of the State-of-the-Science of Endocrine Disruptors (WHO-IPCS 2002a). This report proposes a collective weight of evidence approach to evaluating whether observed effects resulting from exposure to EDCs can be attributed to an endocrine-mediated mechanism. The report presents several examples where the mechanism of action for selected chemicals is clearly related to direct perturbations of endocrine function and ultimately to adverse in vivo effects. It also explores available human health data and draws on relevant data from studies with laboratory animals to assess the biological plausibility of adverse effects on human health.

Endocrine Disrupting Compounds: Priority List of EDCs (GWRC 2003). The Global Water Research Coalition (GWRC) developed a preliminary list of compounds considered to be a priority for credible analytical determination of EDCs in water. The GWRC list was prepared on the basis of information received from the members of GWRC and submitted for further refinement to the workshop of GWRC members held in South Africa (October 2002). A number of criteria were considered for inclusion of a chemical in the list, including possession of EDC properties, persistence, and concern to at least two of the GWRC members. The EDC priority list addresses compounds which have been reported to interact with hormones and receptors associated with the reproductive system, particularly estrogens.

Additional sources of information were considered but had limited utility for identifying in vivo data suitable for use as a point of departure in risk assessment. These sources include:

EDSP, Reference Chemicals Used in EDSP Tier 1 Prevalidation Studies (December 10, 2003) (EDSP 2003). The U.S. EPA provided the following description of the chemicals included in the EDSP Tier 1 Prevalidation Studies:

The chemicals contained in this document were selected as reference chemicals for prevalidation studies in the EDSP assay validation program because they produced a well-documented positive response in one or more Tier 1 assays by an identified mode of action. Some chemicals may act by more than one mode of action. It should be noted that this is **not** intended to be a list of endocrine disruptors, and it should not be concluded a chemical on this list is necessarily an endocrine disruptor since that designation also requires an association with an adverse effect identified in a Tier 2 study. Although some chemicals on this list have been studied extensively, others have not been studied in Tier 2 assays.

For the most part, chemicals used in tests in the general literature have not been included in these tables as such studies raise questions about comparability of protocols and data. Thus, the entries in the tables represent studies conducted in the EDSP validation program, EPA ORD laboratories, and the OECD assay validation program. The exceptions to this are the studies shown for the adult or intact male assay, which was developed by the US chemical industry and tested in various industry laboratories.

Endocrine Disruptor Knowledge Base (EDKB). The Federal Food and Drug Administration's (FDA's) Endocrine Disruptor Knowledge Base (EDKB) is an online database containing experimental bioassay data for approximately 2,000 natural and synthetic compounds (FDA-EDKB 2007). Currently, data compiled in the database are limited to bioassay data from in vitro or short-term in vivo bioassays that indicate estrogenic or androgenic activity, including in vitro estrogen receptor competitive binding assays, in vitro cell proliferation assays, in vitro reporter gene assays, and in vivo uterotrophic assays (i.e., assays based on uterine weight gain and vaginal cornification) (FDA-EDKB 2007). These data are useful for screening purposes and to indicate mechanism of action but generally are not suitable for use as a point of departure for risk assessment, with the possible exception of data from uterotrophic assays conducted with intact immature rodents.

Criterion 2: Occurrence and Exposure

Preference was given to potential EDCs that are known to occur or that might be expected to occur in drinking water, particularly in the U.S. Criteria for selection on this basis are listed in order of decreasing preference:

- Previously reported to occur in drinking water, particularly in the U.S. or other developed nations,
- Likely to be found in drinking water based on known occurrence in source waters for drinking water treatment plants and known or predicted resistance to conventional drinking water treatment,
- Likely to be found in drinking water based on known sources and likely pathways to raw water and known or predicted resistance to conventional drinking water treatment, and can be detected and quantified in drinking water with a currently available analytical method.

Chemicals might be expected to resist conventional drinking water treatment based on experimental treatment data (bench-, pilot-, or full-scale), physical-chemical properties, and/or environmental fate data. The sufficiency of the database on occurrence for each candidate contaminant was evaluated. For some EDCs detected in drinking water, there are few data to describe the frequency of detection or variation in concentrations among drinking water treatment supplies or even within a single supply.

Bioavailability also affects uptake and internal exposure to contaminants. Chemicals with molecular masses up to almost 1,000 Daltons have been shown to be bioavailable and to be transferred across biological membranes. Most potential EDCs have masses in the range of 200-600 Daltons. U.S. EPA considers certain substances including most polymers with average molecular weight greater than 1,000 Daltons (atomic mass units) to have a low potential to cause endocrine disruption (USEPA 2005). No polymers were included among the initial candidate chemicals.

Criterion 3: Effects

Adverse effects are considered to be important for regulation of chemicals in drinking water (Fisher 2004) as well as for risk assessment. Criteria for determining which endocrine-mediated effects can be characterized as adverse has been a major point of contention among various experts, government agencies, environmental advocacy groups, and industry/ trade groups. In general, short-term changes in subclinical indicators like enzyme activities and hormone levels are not considered to be adverse, since the body possesses compensatory mechanisms that can adjust these back into the normal range. However, some proponents of the precautionary principle argue that any change that requires an organism to respond by repairing damage or modulating hormone levels might result in diversion of energy from other critical bodily processes like reproduction or impair the ability of an organism to compensate for other stressors. Also, a change that has no effect in an adult animal might have a severe and/or irreversible effect on a developing animal at critical periods in its development. Conversely, an effect observed in a developing organism might not persist into adulthood. Thus, bioassays that capture reproduction, development, and maturation are most likely to identify persistent adverse effects (Jobling et al. 1998).

Descriptions of severity of effects, potency, and pharmacokinetics (e.g., half-life, bioaccumulation) influenced the selection of target EDCs for this study. Consideration was also given to availability of high quality studies resulting in data that are useful for conducting human health risk assessments, including human epidemiology, occupational health, or clinical studies and studies of endocrine disruptive effects in intact laboratory animals that serve as surrogates for humans.

Criterion 4: Mode of Action

The initial list of candidate EDCs included chemicals that have been reported to have the potential to act through each of the EAT modes of action (estrogenic/ anti-estrogenic, androgenic/ anti-androgenic, and thyroid-related). Only these classes of EDCs and modes of action were considered, and an emphasis was placed on the estrogenic EDCs because they generally are the best characterized group.

Criterion 5: Interest in Specific Contaminants

During the process of finalizing the list of EDCs selected for development of acceptable daily intake (ADIs) and analysis in water, interest in specific contaminants among the project advisory committee (PAC), utilities, the public, and the scientific community in general was evaluated. Utilities or members of the PAC expressed interest in the following chemicals: styrenes, polybrominated diphenyl ethers (PBDEs), polychlorinated biphenyls (PCBs), phthalates, and resorcinol. Contaminants currently being screened by the EDSP were considered for inclusion in the initial list of candidate EDCs. California Department of Health Services (DHS) specified in its Draft Groundwater Recharge Reuse Regulations that certain hormones, EDCs, pharmaceuticals, and other contaminants that serve as indicators of wastewater presence will be monitored annually in recycled water used to recharge a groundwater basin designated for use as a domestic water supply (Table 2.3).

Table 2.3
Emerging contaminants specified for monitoring under California DHS Draft Groundwater Recharge Reuse Regulations

Contaminant Class	Contaminants
Hormones	Ethynylestradiol
	17ß-Estradiol
	Estrone
Industrial EDCs	Bisphenol-A
	Nonylphenol
	Nonylphenol polyethoxylate
	Octylphenol
	Octylphenol polyethoxylate
	Polybrominated diphenyl ethers
Pharmaceuticals and other substances	Acetaminophen
	Amoxicillin
	Azithromycin
	Caffeine
	Carbamazepine
	Ciprofloxacin
	Ethylenediamine tetra-acetic acid (EDTA)
	Gemfibrozil
	Ibuprofen
	Iodinated contrast media
	Lipitor
	Methadone
	Morphine
	Salicylic acid
	Triclosan

Source: Ca DHS 2003

SELECTION OF TARGET ANALYTES – PHYTOESTROGENS

Phytoestrogens are naturally occurring compounds present in plants, such as soybeans, or plant products, such as fruits, that can exhibit estrogenic effects in the body due to their similarity to hormone steroids. Several decades ago a positive relation between cow mastitis and phytoestrogens: formononetin-like and genistein-like substances was discovered (Brookbanks, Welch, and Coup 1969a). Since then, many biological effects in animals and humans, that can be associated with phytoestrogens, have been discovered. Most of these effects are found beneficial due to the anticarcinogenic, antiatherogenic and antioxidative potentials of phytoestrogens (Adlercreutz et al. 1991, Wu et al. 1996, Ingram et al. 1997, Murkies et al. 2000, Lu et al. 2001, Shu et al. 2001). However, some controversial results are reported indicating the complexity of this issue.

Overall, phytoestrogens have much weaker estrogenic effects when compared to estrogens. But they constitute a constant daily intake at relative high levels and can be a cause for various observed human effects. The nature of the effects may depend on factors such as age, exposure dose, differences in compounds and their interactions, strains, individual's susceptibility, presence of other unknown compounds with estrogenic activity, and other yet unknown factors.

Based upon a review of available literature 15 common dietary phytoestrogens were identified. These phytoestrogens and their major sources are summarized in Table 2.4. Information on phytoestrogen occurrence in various matrices, daily dietary intake and estrogenic potency were also sought for the selection process. The widely occurring genisten and daidzein were two most intensively studied compounds and were included in most studies, while the inclusion of other compounds varied dramatically in reviewed literature. The occurrence of some phytoestrogens metabolites were reported in human blood serum and urine (Zheng et al. 1999, Morton et al. 2002, Grace et al. 2003b, Grace et al. 2003a, Valentin-Blasini et al. 2003). Of these, only equol was selected for this study due to its relatively high occurrence. To evaluate the estrogenic potency of the 15 phytoestrogens in the final list, bioassay experiments were used to evaluate in vitro estrogenicity. The experiments were designed based on literature data in combination with bioactivity predicted with Schrödinger, a commercial software that can predict the binding potential of each phytoestrogen with selected crystallized structure of human estrogen receptor α, 1QKU. E-screen results obtained with MCF-7 human breast cancer cells (Table 2.5) are in general agreement with literature data, but some values are lower than reported data suggesting the higher sensitivity of the MCF-7 in vitro method.

Our results have shown that most phytoestrogens have very weak estrogenicity when compared to mammalian steroid hormones such as 17β-estradiol that has an EC_{50} of 6 pico Molar (pM). By considering the daily dietary intake, all phytoestrogens with EC_{50} below 3 μM are included in the final list of phytoestrogens for evaluation. This criterion potentially eliminated catechin, chrysin, enterolactone, glycitein, kaempferol, matairesnol and quercetin. Chrysin, glycitein and matairesnol were left in the final list as they showed promising analytical method results and were believed to have relatively high occurrence. Kaempferol and quercetin were eliminated due to analytical challenges. They have consistently shown poor recoveries (<30%) during solid-phase extration. A comprehensive evaluation of these estrogenic potency data obtained for phytoestrogen standards using the E-screen bioassay, together with their occurrence and daily intake obtained from the literature eventually resulted in our final list of eleven phytoestrogens, which are presented in Figure 2.1.

Table 2.4
Common dietary phytoestrogens

Phytoestrogens	Source
Apigenin	Parsley, artichoke, basil, celery, and other leafy plants
Biochanin A	Legumes (i.e., soy beans) and red clover
Catechin	Green tea
Coumestrol	Alfalfa and forage crops
Chrysin	Plant Passilora Caerula
Daidzein	Legumes (i.e., soy beans) and red clover
Enterodiol	From Secoisolariciresinol due to intestinal bacterial action
Enterolactone	From matairesinol due to intestinal bacterial action
Equol	Metabolite from daidzein
Formononetin	Clover (including red clover)
Genistein	Legumes (i.e., soy beans) and red clover
Kaempferol	Fruits, vegetables, and tea
Matairesinol	Oilseeds such as sesame (Thompson et al. 1991)
Naringenin	Citris fruits and tomatoes
Quercetin	Fruits and vegetables

Table 2.5
Experimentally obtained in vitro EC50 in comparison with the literature data

Title	Literature in vitro EC50 (M)	E-screen (M)
17β-estradiol*	4.5E-12 ~ 9E-10	5.9E-12
Apigenin	4.1E-7 ~ 1E-5	1.5E-06
BiochaninA	2.1E-05	1.8E-07
Catechin	nonbinder	>4.3E-5
Chrysin	nonbinder	3.8E-06
Coumestrol	4E-8 ~ 1E-6	9.6E-09
Daidzein	6.76E-7 ~ 1E-4	2.3E-07
Enterolactone		4.4E-06
Equol	6E-7 ~ 1E-5	4.9E-08
Formononetin	slight binder	7.7E-07
Genistein	9.7E-9 ~ 4.6E-7	4.7E-08
Glycitein		1.1E-04
Kaempferol	8.3E-7 ~ 6.86E-5	3.4E-06
Matairesinol		>> 5.0E-6
Naringenin	1E-5 ~ 3E-5	5.8E-07
Quercetin	1.3E-05	3.3E-05

* 17ß-estradiol is included as the standard estrogen for comparison of potencies.

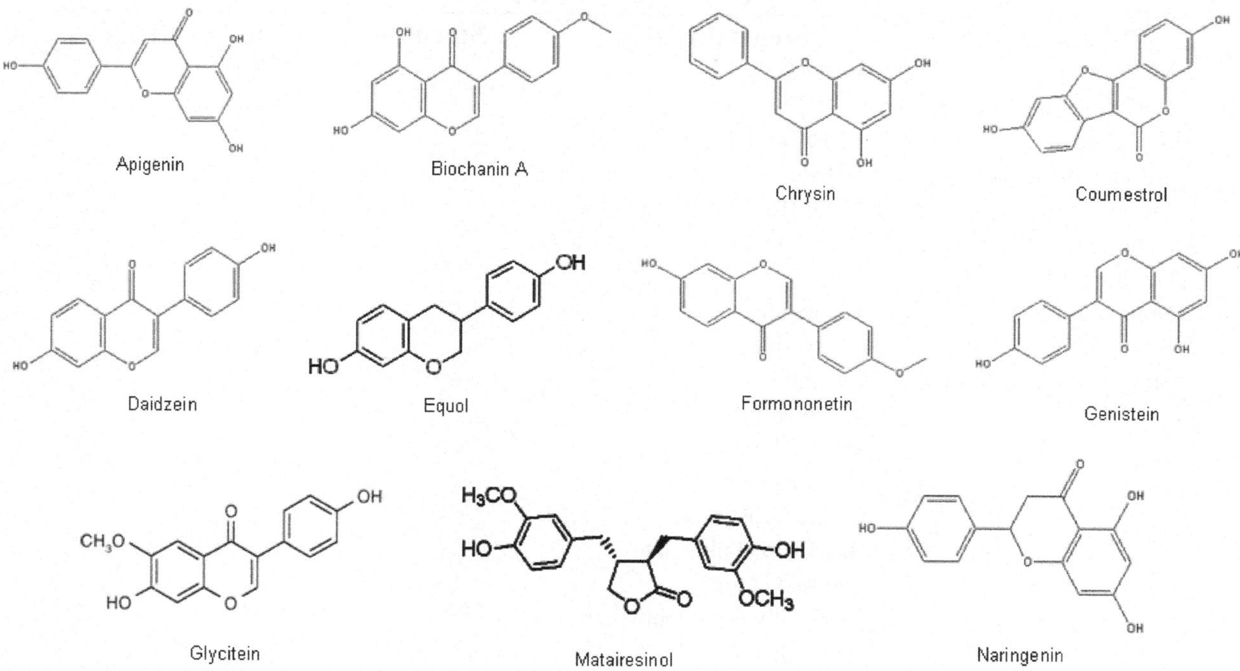

Figure 2.1 Structures of phytoestrogens chosen for final evaluation

FINAL SELECTION OF TARGET ANALYTES

Based on the above mentioned criteria for pharmaceuticals, potential EDCs and phytoestrogens, the final compounds, or analytes chosen for this study are listed in Table 2.6. Method details, including reporting limits, extracton recoveries and instrumental parameters for target analytes are located in Appendix A.

Table 2.6
Final Selection of Target Analytes

Pharmaceuticals	Potential EDCs	Steroids	Phytoestrogens
Atenolol	Atrazine	Estradiol	Genistein
Atorvastatin	Linuron	Estrone	Daidzein
o-Hydroxy atorvastatin	Bisphenol A	Ethynylestradiol	Formononetin
p-Hydroxy atorvastatin	Triclosan	Progesterone	Biochanin A
Carbamazepine	BHA	Testosterone	Apigenin
Diazepam	BHT		Naringenin
Phenytoin	DEET		Coumestrol
Enalapril	Octylphenol		Chrysin
Fluoxetine	Benzophenone		Matairesinol

(continued)

Table 2.6, Continued

Pharmaceuticals	Potential EDCs	Steroids	Phytoestrogens
Meprobamate	α-BHC		Equol
Norfluoxetine	β-BHC		Glycitein
Risperidone	γ-BHC (Lindane)		
Sulfamethoxazole	TCEP		
Trimethoprim	TCPP (Fyrol PCF)		
Diclofenac	Diazinon		
Gemfibrozil	δ-BHC		
Naproxen	Traseolide		
Simvastatin	Galaxolide		
Simvastatin hydroxy acid	Tonalide		
	Vinclozolin		
	Metolachlor		
	Musk Ketone		
	Octachlorostyrene		
	Butylbenzyl phthalate		
	Methoxychlor		
	Bis(2-ethylhexyl)phthalate		
	Nonylphenol		

CHAPTER 3
OCURRENCE OF TARGET COMPOUNDS

SUMMARY OF PARTICIPATING UTILITIES

More than 300 samples were collected from 20 drinking water treatment plants (DWTPs) and four wastewater treatment plants (WWTPs) across the continental U.S. Of these samples, 195 were from states in the Pacific Time Zone, 43 from states in the Central Time Zone, and 88 from states in the Eastern Time Zone (Figure 3.1). All treatment plants were labeled numerically to ensure their anonymity (Table 3.1).

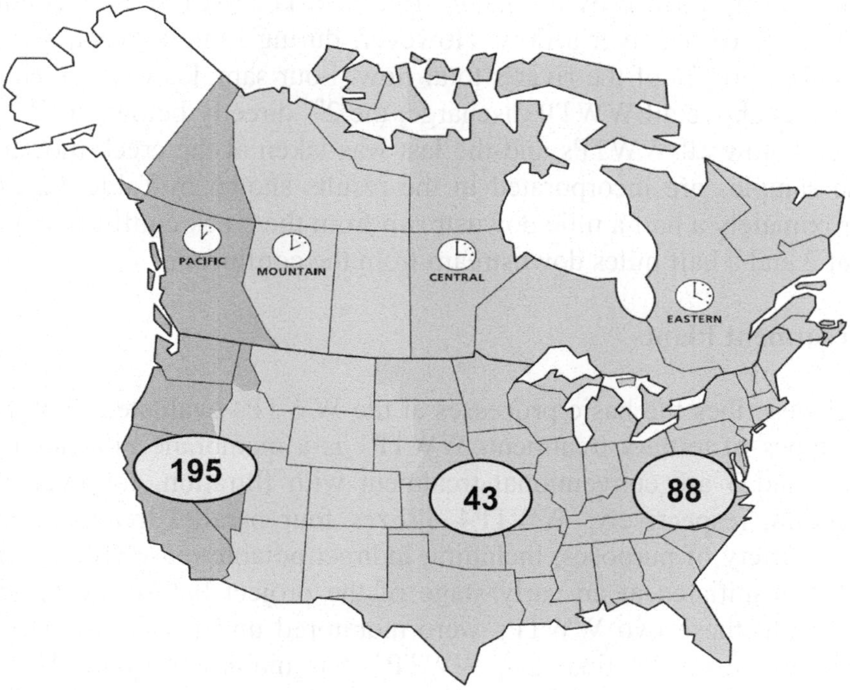

Figure 3.1 Number of samples per time zone

Drinking Water Treatment Plants

Of the DWTPs evaluated, two utilized only groundwater as source water, while all others utilized surface waters. The facilities evaluated generally used conventional treatment techniques consisting of coagulation, flocculation, sedimentation, primary disinfection, filtration and secondary disinfection (for the distribution system); however, there are some variations among these treatment plants. For instance, the application of preoxidation and the type of preoxidants differ. Also, chlorine may be added at various locations in the treatment process in addition to the secondary disinfection for maintaining disinfectant residual. Six of the facilities evaluated used ozonation as the primary disinfectant. The treatment processes summarized in Table 3.1 indicate that chloramine is the most common secondary disinfectant among the treatment plants evaluated.

Raw and finished drinking water samples were collected from most DWTPs. Additionally, one or two distribution sites were sampled at 10 of the participating utilities with a preference towards collection points with longer retention times in order to determine whether target analyte concentrations change under actual distribution system conditions. Three of the utilities in this study were sampled at least twice at different times of the year to evaluate seasonal variations.

Utilities DWTP1 through DWTP9 have known wastewater impact to their source water, with wastewater comprising from 1 to 17% of the total water supply. DWTP1 withdraws water from a lake that receives water from a WWTP and two tributary lakes, each of which contain less than 8% of effluent from an upstream WWTP. DWTP5 withdraws water from a reservoir that also receives effluent from a tertiary WWTP. The reservoir was sampled at three locations which influence the drinking water intake. DWTP7 and DWTP8 share the same source water. DWTP7 is located upstream of DWTP8 along a common river, which receives water from tributaries and a creek; the latter mostly formed by discharge from WWTPs. At low flow conditions, the creek contributes up to 27% of the river's flow. However, during sample events for this project, the creek made up only 2 to 6% of the river's total flow. Four samples were taken along the creek. The first sample was above all WWTP discharge, the 2nd directly below the 2nd WWTP. The 3rd sample was taken below all WWTPs and the last was taken at the creek mouth before it enters the river. These samples are incorporated in the results shown in Table 3.2. The intake from DWTP7 is approximately a half a mile downstream from the creek confluence on the river, while DWTP8 is about 2 and a half miles downstream from the confluence.

Wastewater Treatment Plants

Table 3.1 describes the basic processes at the WWTPs evaluated. WWTPs1, 2 and 3 all utilize different types of tertiary treatment. WWTP1 is a membrane bioreactor (MBR) facility, while WWTPs 2 and 3 are conventional treatment with filtration, followed by UV and free chlorine disinfection, respectively. WWTP4 utilizes four parallel treatment trains to provide reuse water for a variety of purposes, including indirect potable reuse (IPR). The monitoring of WWTP1 and 2 was initiated at an early stage of the project before the target analytes were finalized. As a result, these two WWTPs were monitored under a preliminary compound list, which differs slightly from the final list. WWTP1 was monitored quarterly for one year and WWTP3 was monitored twice (May and August) for seasonal evaluation. In addition to influent and effluent samples for all WWTPs, WWTP3 and WWTP4 also were sampled between unit treatment processes to understand the fate of target compounds through each stage of treatment. Although monitoring of the WWTPs yielded interesting results, these data were not pertinent to the risk evaluation from drinking water and therefore results for all 4 WWTPs are included only in Appendix B and not used for any risk assessments.

Table 3.1
Summary of participating water utilities

Water treatment plants evaluated	Treatment Processes (in order)
DWTP1	Coagulation/flocculation, sedimentation, free chlorine, filtration, chloramine
DWTP2	Chlorine dioxide preoxidation, coagulation/flocculation, sedimentation, filtration, UV, free chlorine
DWTP3	Chlorine dioxide preoxidation, coagulation/flocculation, sedimentation, filtration, UV, free chlorine
DWTP4	Free chlorine preoxidation, coagulation/flocculation, sedimentation, free chlorine, filtration, and free chlorine
DWTP5	Free chlorine preoxidation, ozone, coagulation/flocculation, filtration, free chlorine
DWTP6	Free chlorine preoxidation, ozone, coagulation/flocculation, filtration, free chlorine
DWTP7	Free chlorine preoxidation, coagulation/flocculation, sedimentation, free chlorine, filtration, and chloramine
DWTP8	Free chlorine preoxidation, coagulation/flocculation, sedimentation, free chlorine, filtration, and chloramine
DWTP9	Free chlorine preoxidation, coagulation/flocculation, sedimentation, free chlorine, filtration, and chloramine
DWTP10	Free chlorine preoxidation, coagulation/flocculation, sedimentation, filtration, chloramine
DWTP11	Free chlorine preoxidation, coagulation/flocculation, sedimentation, filtration, chloramine
DWTP12	Free chlorine, chloramine
DWTP13	Free chlorine preoxidation, ozonation, coagulation/flocculation, sedimentation, chloramine
DWTP14	Filtration, ozone, chloramine
DWTP15	Free chlorine preoxidation, coagulation/flocculation, sedimentation, free chlorine, filtration, and chloramine
DWTP16	Coagulation, flocculation, sedimentation, ozone, dual media filtration, chloramine,
DWTP17	Free chlorine preoxidation, coagulation/flocculation, sedimentation, free chlorine, filtration, and chloramine
DWTP18	Free chlorine only
DWTP19	Ozone preoxidation and intermediate oxidation, coagulation/floccualtion, sedimentation, GAC/sand biofiltration, chloramine
DWTP20	Ozone preoxidation and intermediate oxidation, coagulation/floccualtion, sedimentation, GAC/sand biofiltration, chloramine
WWTP1	Primary, secondary, MBR
WWTP2	Primary, secondary, filtration, UV
WWTP3	Primary, secondary, filtration, free chlorine
WWTP4	Four trains for water reuse, source water is secondary wastewater effluent ▪ Mono-media filtration, free chlorine ▪ Microfiltration and RO ▪ Microfiltration and two stage RO ▪ Microfiltration, RO, UV-AOP

METHOD SUMMARY

A brief summary of the analytical methods developed and employed for this project is presented in this section. Additional details are provided in Appendix A, including extraction methods, QA/QC data (blanks, recoveries, matrix spike recoveries), and analytical instrument parameters. The methodologies employed have been published in peer-reviewed literature previously; however, for this project a few modifications have been made and are noted in the following sections (Vanderford et al. 2003, Trenholm et al. 2006).

Sample Collection

For each sample, 2 liters of water were collected in pre-silanized, amber glass bottles preserved with sodium azide and ascorbic acid. In most cases, samples were collected based upon hydraulic retention times following plug flow calculations. Water quality measurements were collected at the same time as the samples by personnel from each utility. Every effort was made to follow the same plug of water through the treatment train. This allows an objective evaluation of the same plug of water in terms of the removal efficiency. For each plant, approximately two sites were selected for triplicate samples, usually the raw source water and finished water, as a quality control measure. This allowed for duplicate samples as well as a matrix spike sample. Additionally, travel/field blanks were collected with each sampling location. All samples were shipped in ice packed coolers overnight and stored at 4 °C until extraction. Laboratory deionized water (DI - MilliQ reagent water) blanks and DI spiked samples were extracted alongside every batch of samples to ensure the integrity of the SPE instrument and analytical process.

Organic Contaminants Monitored and Analytical Methods

Pharmaceuticals, potential endocrine disrupting compounds (EDCs), steroid hormones, and phytoestrogens were selected as the four primary groups of compounds monitored in this study. Sixteen pharmaceuticals and metabolites of atorvastatin, fluoxetine and simvastatin were selected based on their toxicity, occurrence potential, representation of drug groups, and prescription volume. Twenty-six potential EDCs and five steroid hormones were included in the compound list due to their potency in generating estrogenic effects. For phytoestrogens, although dietary intake is considered a more significant pathway in comparison to drinking water consumption, their occurrence in finished water and fate in the aquatic systems remains unknown. The inclusion of phytoestrogens provided a comparison of water intake with dietary intake in term of estrogenic exposure.

Three separate automated solid phase extraction (ASPE) methods were employed for the final target compound list and analyzed by LC-MS/MS or GC-MS/MS. Each extraction used 500 mL of sample, except for some WWTP samples, in which only 50 to 100 mL volume was used to avoid potential damage to sensitive analytical equipment. In the case of WWTP and source water samples that appear turbid, samples were filtered using GF/F glass fiber filters prior to ASPE. All extraction methods were performed using an Autotrace™ ASPE system using glass, 200 mg, 5 cc, HLB cartridges (Oasis, Waters Corporation). Each method followed a similar strategy in which the SPE cartridges were conditioned, samples passed through, the cartridge rinsed, dried with nitrogen gas, and analytes eluted into 15 mL vials. After elution, extracts were concentrated using high-purity nitrogen and kept at -20 °C until instrumental analysis.

The ASPE used for the pharmaceutical method, also included three potential EDCs (atrazine, linuron, and bisphenol A). Each sample was spiked with a stock solution of isotopically labeled standards prior to extraction for isotope dilution quantitation. After sample loading, the SPE cartridge was rinsed with DI water, dried with nitrogen, and then eluted using a combination of MTBE and methanol. Extracts were concentrated to 500 µL using high-purity nitrogen in preparation for LC-MS/MS analysis.

The second ASPE method was used for the extraction of steroid hormones and phytoestrogens. Since these groups share similar in structures, one extraction method could be employed for both groups. Prior to extraction, each sample was spiked with isotopically labeled standards for isotope dilution and surrogate standards. Because only two phytoestrogens had isotopically labeled versions available, isotope dilution could not be performed for this group of compounds. The two available isotopes (genistein-d_4 and daidzein-d_3) were used as surrogate standards to monitor the extraction process; however, data were not adjusted for surrogate recovery. Due to the compounds' strong affinity to the HLB cartridge, a stronger rinse was applied to help minimize matrix interference. After sample loading, each cartridge was rinsed with a methanol solution. The SPE cartridges were then dried with nitrogen and eluted with a combination of MTBE and methanol. Extracts were concentrated using nitrogen and split into separate fractions. The first fraction (phytoestrogens) went directly to LC-MS/MS analysis, while the second fraction (steroids) was further concentrated to 250 µL before LC-MS/MS analysis.

The third ASPE method was used for the remaining potential EDCs. Each sample was spiked with isotopically-labeled standards for isotope dilution and with surrogate standards prior to extraction. Not all the compounds in this group had available isotopes, so surrogates were used in some cases to monitor the extraction process. After sample loading, SPE cartridges were rinsed with DI water, dried with nitrogen, and eluted with a combination of dichloromethane, MTBE, and methanol. Extracts were concentrated and solvent transferred to iso-octane, at which time internal standards were added prior to GC-MS/MS analysis. Nonylphenol represents a technical mixture of nonylphenol isomers and results are based on the total summation of these isomers. Although octylphenol can also exist as different isomers, 4-*tert*-octylphenol is by far the most common in the environment and was used for this study, and is simply referred to as octylphenol.

OCCURRENCE OF EDCS AND PPCPS IN WATER

For this study, some sites were sampled numerous times (as replicates and/or as temporal sampling events), while other sites were only sampled once. When comparing sites it is important to use representative information, so as not to over emphasize one site or another, simply due to sampling frequency. Therefore, all sites with multiple samples were averaged and are comparable to other sites where single sampling occurred. For example, a maximum value shown in the data tables may represent a maximum value based on the average of multiple samples for one site. However, for the drinking water equivalent level (DWEL) calculations (provided in Chapters 6 and 7), the single highest discrete sample concentration was used as a conservative "worst-case" scenario approach. Therefore, the maximum values reported for finished drinking water in Table 3.4 may differ slightly from the highest detected value used in

Chapters 6 and 7 for risk assessments. Results are also shown as a comparison of two different approaches to dealing with non-detected values when averaging results. The results shown on the left of Tables 3.2 to 3.5 represent calculations where only detected values are averaged and non-detected values are ignored. The results on the right show the same data; however, instead of ignoring non-detects, half the reporting limit (RL) was used when calculating the average. For some compounds an analyte was detected just above the reporting limit in one sample, but not in a replicate sample. In these cases, the final averaged result is below the reporting limit and therefore not reported. The more conservative approach is to ignore the non-detected values while averaging data. However, the argument could be made that this approach may also over emphasize compound detection. These examples demonstrate the importance of how handling non-detected values, especially while averaging, can affect data reporting.

Occurrence in Source Water Samples

Table 3.2 shows results for source waters along with frequency of detection, maximum concentration, and median detected concentration where n equals number of sites (as opposed to the number samples). These are not necessarily the raw water collected from the DWTPs, but rather from the waters that impact the raw water intakes/wells. Source water samples were generally collected upstream of raw water intakes for DWTPs, except for groundwater systems where monitoring wells were evaluated. Samples were collected from 12 sites in the Pacific, 2 sites in the Central and 4 sites in the Eastern Time Zone. Results show a relatively high frequency of occurrence of target compounds, with some compounds detected in μg/L concentrations such as atrazine, TCEP, and TCPP. Eight of the pharmaceuticals were consistently detected in over 50% of the sites with sulfamethoxazole having the highest observed concentration at 890 ng/L. Phytoestrogens were detected in less than half the sites with concentrations of 25 ng/L or less. Steroid hormones (estrone, progesterone and estradiol) were detected at several sites with the highest level detected at 6.2 ng/L for estrone.

In comparison to raw water occurrence results (Table 3.3), many compounds detected in source waters were not detected in raw waters. This could be attributed to dilution and further degradation of the analytes due to proximity of intake structures to source water sampling locations. Some compounds that were detected in both source and raw water exhibited lower concentrations in the raw water. For instance, estrone was detected at a high of 6.2 ng/L in source waters, but had a max detection of only 0.94 ng/L in the raw water and was not detected at all in finished drinking water. This anomally also could be caused by stratification and flow pattern differences between source water sampling points and raw water intake (or well) location.

Table 3.2
Results for source water (n=18)

Non-detected results ignored					Non-detected results averaged using ½ RL				
	Max	Median	Frequency			Max	Median	Frequency	
Compound	ng/L	ng/L	#	%	Compound	ng/L	ng/L	#	%
Sulfamethoxazole	890	12	16	89	Carbamazepine	430	6.7	15	83
Carbamazepine	430	11	15	83	Sulfamethoxazole	890	12	15	83
Atrazine	1700	1.6	14	78	Atrazine	1700	1.6	14	78
Phenytoin	180	37	13	72	Phenytoin	180	37	13	72
Meprobamate	560	25	13	72	Meprobamate	560	25	13	72
Atenolol	560	27	12	67	TCEP	1800	180	12	67
TCEP	1800	200	12	67	Trimethoprim	78	9.1	12	67
Trimethoprim	78	9.1	12	67	Atenolol	560	47	11	61
Estrone	6.2	1.2	11	61	DEET	450	180	10	56
DEET	450	180	10	56	Diclofenac	26	5.4	10	56
Diclofenac	26	5.4	10	56	Gemfibrozil	52	17	10	56
Gemfibrozil	52	17	10	56	Triclosan	49	7.3	10	56
Triclosan	49	7.3	10	56	Estrone	6.2	2.9	9	50
Nonylphenol	630	110	9	50	TCPP (Fyrol PCF)	1600	510	9	50
TCPP (Fyrol PCF)	1600	510	9	50	Bisphenol A	120	22	8	44
Apigenin	4.2	2.2	8	44	Glycitein	25	8.3	8	44
Bisphenol A	120	26	8	44	Nonylphenol	630	140	8	44
Fluoxetine	27	3.4	8	44	Apigenin	4.2	2.4	7	39
Glycitein	25	8.3	8	44	Diazepam	2.7	1.4	7	39
Naproxen	150	23	8	44	Fluoxetine	27	5.1	7	39
Diazepam	2.7	1.4	7	39	Naproxen	150	31	7	39
Galaxolide	990	320	7	39	Naringenin	8.8	1.8	7	39
Naringenin	17	2.5	7	39	Atorvastatin	16	3.3	6	33
Atorvastatin	16	3.3	6	33	Galaxolide	990	360	6	33
Benzophenone	510	55	6	33	Norfluoxetine	3.7	1.3	6	33
Genistein	9.5	2.7	6	33	o-Hydroxy atorvastatin	16	3.1	6	33
Norfluoxetine	3.7	1.3	6	33	p-Hydroxy atorvastatin	26	3.7	6	33
o-Hydroxy atorvastatin	16	3.1	6	33	Simvastatin hydroxy acid	1.4	0.43	5	28
p-Hydroxy atorvastatin	26	3.7	6	33	Tonalide	220	63	5	28
Daidzein	9.2	3.2	5	28	Benzophenone	510	59	4	22
Formononetin	22	6.4	5	28	Daidzein	7.3	3.4	4	22
Octylphenol	68	34	5	28	Genistein	8.4	4.3	4	22

(continued)

Table 3.2 (Continued)

Non-detected results ignored					Non-detected results averaged using ½ RL				
	Max	Median	Frequency			Max	Median	Frequency	
Compound	ng/L	ng/L	#	%	Compound	ng/L	ng/L	#	%
Simvastatin hydroxy acid	1.4	0.43	5	28	Linuron	1.5	1.0	4	22
Tonalide	220	63	5	28	Octylphenol	40	28	4	22
Chrysin	1.4	1.2	4	22	Chrysin	1.4	1.2	3	17
Linuron	2.9	2.5	4	22	Formononetin	22	7.7	3	17
Metolachlor	280	140	4	22	Metolachlor	99	89	3	17
Musk Ketone	85	42	4	22	Musk Ketone	85	42	3	17
Progesterone	3.1	1.7	4	22	Progesterone	1.7	1.2	3	17
Biochanin A	8.3	5.5	3	17	Biochanin A	4.4	3.3	2	11
Butylbenzyl phthalate	60	55	3	17	BHA	31	31	1	6
BHT	49	45	2	11	BHT	41	41	1	6
Enalapril	0.38	0.36	2	11	Butylbenzyl phthalate	55	55	1	6
Estradiol	2.0	1.4	2	11	Coumestrol	1.0	1.0	1	6
BHA	50	50	1	6	Bis(2-ethylhexyl) phthalate	280	280	1	6
Coumestrol	2.0	2.0	1	6	Enalapril	0.30	0.30	1	6
Diadzinon	10	10	1	6	Estradiol	0.83	0.83	1	6
Bis(2-ethylhexyl) phthalate	280	280	1	6	Risperidone	3.7	3.7	1	6
Matairesinol	6.3	6.3	1	6	Traseolide	34	34	1	6
Risperidone	3.7	3.7	1	6					
Traseolide	34	34	1	6					

Occurrence in Raw Intake Water Samples

Table 3.3 provides the results for raw (untreated) water samples for 19 sites in the U.S., along with frequency of detection, maximum detected concentration, and median detected concentration. Raw water is defined here as samples taken just prior to entering the DWTP. Not surprisingly, DWTPs with known wastewater impact tend to exhibit higher concentrations and also higher detection frequency of pharmaceuticals and potential EDCs in their raw water samples. Overall, pharmaceuticals are the most frequently detected compounds in raw waters. Of the 16 selected pharmaceuticals, eight were detected in over 50% of the sites evaluated. Meprobamate and sulfamethoxazole were the two most frequently detected pharmaceuticals. The hydroxylated metabolites of atorvastatin co-occurred in every case where the parent compound, atorvastatin, also was detected. The other metabolite norfluoxetine was not detected despite the occurrence of fluoxetine. Simvastatin and the simvastatin hydroxy acid metabolite also were not detected in any raw water samples.

Both steroid hormones and phytoestrogens had low frequency of occurrence in raw water. The concentrations of steroid hormones were generally below 5 ng/L, with the exception of estradiol which was detected at 17 ng/L at one site. This single detection for estradiol is difficult to explain and there was insufficient sample volume for reanalysis. Also, this site was not evaluated for cellular estrogenicity, so no corresponding bioassay data are available. In the potential EDCs group, atrazine, TCEP, TCPP, DEET, and galaxolide were detected in raw water from DWTPs with higher wastewater impact. Approximately half of the potential EDC target analytes were not detected in any sample. Results for Table 3.3 are shown using both data analysis approaches for calculating non-detects (similar to Table 3.2).

Table 3.3
Results for raw intake water (n=19, *n=17)

Non-detected results ignored					Non-detected results averaged using ½ RL				
Compound	Max ng/L	Median ng/L	Frequency #	Frequency %	Compound	Max ng/L	Median ng/L	Frequency #	Frequency %
Sulfamethoxazole	110	12	17	89	Sulfamethoxazole	110	12	17	89
Meprobamate	73	8.2	16	84	Meprobamate	73	8.2	16	84
Atrazine	870	32	15	79	Atrazine	870	32	15	79
Carbamazepine	51	4.1	15	79	Carbamazepine	51	4.1	14	74
Estrone	0.94	0.33	15	79	Estrone	0.94	0.33	14	74
Phenytoin	29	5.0	14	74	Phenytoin	29	4.6	13	68
Atenolol	36	2.2	12	63	Atenolol	36	1.6	12	63
Gemfibrozil	24	2.2	11	58	Gemfibrozil	24	2.2	11	58
Naproxen	32	0.93	11	58	Naproxen	32	0.93	11	58
Trimethoprim	11	0.75	11	58	TCEP	530	110	10	53
TCEP	530	120	10	53	Trimethoprim	11	1.9	9	47
TCPP (Fyrol PCF)	720	180	8	42	TCPP (Fyrol PCF)	720	180	8	42
Nonylphenol	130	100	8	42	Metolachlor	55	17	7	37
Metolachlor	81	17	7	37	DEET	110	82	6	32
Apigenin*	3.0	1.8	6	35	Triclosan	6.4	2.8	5	26
DEET	110	85	6	32	Apigenin*	3.0	1.0	3	18
Triclosan	6.4	3.0	6	32	Formononetin*	4.3	4.2	3	18
Linuron	9.3	4.1	5	26	Genistein*	3.0	2.0	3	18
Formononetin*	6.1	3.3	4	24	Diclofenac	1.2	0.81	3	16
Genistein*	4.3	2.2	4	24	Galaxolide	48	41	3	16
Diclofenac	1.2	1.1	4	21	Linuron	6.3	4.8	3	16
Galaxolide	48	36	4	21	Nonylphenol	130	110	3	16
Progesterone	3.1	2.1	4	21	Progesterone	2.5	1.7	3	16
Daidzein*	5.6	1.4	3	18	Glycitein*	2.4	2.1	2	12
Glycitein*	3.4	2.3	3	18	Atorvastatin	1.4	1.1	2	11
Atorvastatin	1.4	0.83	3	16	Bisphenol A	11	9.6	2	11
Bisphenol A	14	6.1	3	16	Diazepam	0.35	0.33	2	11
Fluoxetine	3.0	0.80	3	16	Fluoxetine	1.6	1.1	2	11
o-Hydroxy atorvastatin	1.2	0.66	3	16	o-Hydroxy atorvastatin	1.2	0.85	2	11
p-Hydroxy atorvastatin	2.0	1.0	3	16	p-Hydroxy atorvastatin	2.0	1.5	2	11

Continued

Table 3.3 (Continued)

Non-detected results ignored					Non-detected results averaged using ½ RL				
	Max	Median	Frequency			Max	Median	Frequency	
Compound	ng/L	ng/L	#	%	Compound	ng/L	ng/L	#	%
Biochanin A	1.5	1.4	2	12	**Testosterone**	0.59	0.58	2	11
Chrysin*	2.1	1.6	2	12	**Chrysin***	2.1	2.1	1	6
Naringenin*	2.6	1.9	2	12	**Coumestrol***	1.0	1.0	1	6
Bis(2-ethylhexyl) phthalate	54	52	2	11	**Daidzein***	3.9	3.9	1	6
Butylbenzyl phthalate	54	52	2	11	**Naringenin***	1.2	1.2	1	6
Diazepam	0.50	0.43	2	11					
Testosterone	1.2	1.1	2	11					
Coumestrol*	2.0	2.0	1	6					
BHT	49	49	1	5					
Estradiol	17	17	1	5					
Ethynylestradiol	1.4	1.4	1	5					

Occurrence in Finished Drinking Water Samples

Table 3.4 shows target analytes detected in finished drinking water for 18 sites in the U.S., along with frequency of detection, maximum concentration, and median detected concentration. Finished drinking water is defined here as samples taken after final disinfectant exiting the DWTP. Results are also shown as a comparison of two different approaches to dealing with non-detected values when averaging results. Compounds such as fluoxetine, glycitein, BHT, diazepam, and progesterone had averaged concentrations from replicates that were below the MRL and therefore do not appear on Table 3.4 where ½ the RL was used for non-detects. However, when non-detects are not averaged, these compounds were reported.

Despite the high occurrence of sulfamethoxazole in raw drinking water, it was seldom detected in finished water. In contrast, concentrations of meprobamate and phenytoin remained relatively unchanged from source to finished drinking waters, verifying previous reports demonstrating the resilience of these compounds to treatment processes (Snyder et al. 2007a). Steroids and most phytoestrogens were not detected in finished drinking water. Only three phytoestrogens (daidazin, genistein and glycitein) were detected and at concentrations near the method reporting limits. Although a maximum of 870 ng/L of atrazine was present in finished drinking water, this value was from a site that is influenced by agriculture and this value is well below the 3000 ng/L MCL for atrazine (U.S. EPA). Flame retardants were detected in several finished drinking water samples with median concentrations of 120 ng/L for TCEP and 210 ng/L for TCPP, which were similar to the median values observed in raw water (Table 3.3).

Table 3.4
Results for finished drinking water (n=18, *n=16)

Non-detected results ignored					Non-detected results averaged using ½ RL				
	Max	Median	Frequency			Max	Median	Frequency	
Compound	ng/L	ng/L	#	%	Compound	ng/L	ng/L	#	%
Atrazine	870	49	15	83	**Atrazine**	870	49	15	83
Meprobamate	42	5.7	14	78	**Meprobamate**	42	5.2	14	78
Phenytoin	19	6.2	10	56	**Phenytoin**	19	5.9	10	56
Atenolol	18	1.2	8	44	**Atenolol**	18	1.2	8	44
Carbamazepine	18	6.0	8	44	**Carbamazepine**	18	5.6	8	44
Gemfibrozil	2.1	0.48	7	39	**DEET**	83	62	6	33
TCEP	470	120	7	39	**Gemfibrozil**	2.1	1.2	5	28
DEET	93	63	6	33	**Metolachlor**	20	17	5	28
Metolachlor	27	16	6	33	**TCEP**	470	160	5	28
TCPP (Fyrol PCF)	510	210	5	28	**TCPP (Fyrol PCF)**	510	210	5	28
Sulfamethoxazole	3.0	0.39	4	22	**Sulfamethoxazole**	3.0	0.32	4	22
Genistein*	2.9	1.4	3	19	**Linuron**	4.2	4.1	2	11
Daidzein*	5.8	3.8	2	13	**Nonylphenol**	100	97	2	11
Fluoxetine	0.82	0.71	2	11	**Daidzein***	2.3	2.3	1	6
Galaxolide	33	31	2	11	**Genistein***	1.3	1.3	1	6
Linuron	6.2	6.0	2	11	**Bisphenol A**	10	10	1	6
Nonylphenol	100	95	2	11	**Galaxolide**	26	26	1	6
Glycitein*	1.3	1.3	1	6	**Triclosan**	1.2	1.2	1	6
BHT	26	26	1	6					
Bisphenol A	25	25	1	6					
Diazepam	0.33	0.33	1	6					
Progesterone	0.57	0.57	1	6					
Triclosan	1.2	1.2	1	6					

Occurrence in Distribution Water Samples

For 10 DWTPs, sampling was conducted at one or more distribution sites, for a total of 15 distribution system locations. Distribution samples are defined here as samples taken post DWTP and in the distribution system with extended retention time. The results, summarized in Table 3.5, presents the data with both approaches for non-detect samples. When ½ MRL was used to average non-detects, fluoxetine and risperidone were no longer reported. When compared with the overall results for the finished drinking water, many compounds dropped to less than the MRL, including triclosan, glycitein, genistein, linuron, BHT, and galaxolide. This could be attributed to dilution factors or further degradation of the analytes in the distribution system. Compounds such as atrazine, meprobamate, phenytoin, carbamazepine, metolachlor, TCEP, and TCPP did not change significantly in the distribution system. These data show that some target analytes are quite persistent through drinking water treatment and are relatively unchanged through the distribution system.

Table 3.5
Results for distribution water (n=15)

Non-detected results ignored					Non-detected results averaged using ½ RL				
	Max	Median	Frequency			Max	Median	Frequency	
Compound	ng/L	ng/L	#	%	Compound	ng/L	ng/L	#	%
Atrazine	930	1.7	13	87	**Atrazine**	930	49	12	80
Meprobamate	40	5.6	12	80	**Meprobamate**	40	5.6	12	80
Phenytoin	16	3.4	11	73	**Phenytoin**	16	3.6	10	67
Atenolol	0.84	0.47	8	53	**Atenolol**	0.84	0.42	7	47
Carbamazepine	10	6.8	6	40	**Carbamazepine**	10	6.8	6	40
TCEP	200	150	6	40	**TCEP**	200	150	6	40
TCPP (Fyrol PCF)	240	210	6	40	**TCPP (Fyrol PCF)**	240	210	6	40
DEET	63	49	4	27	**DEET**	63	49	4	27
Gemfibrozil	1.2	0.43	4	27	**Gemfibrozil**	1.2	0.43	4	27
Metolachlor	22	18	3	20	**Metolachlor**	22	18	3	20
Naringenin	24	3.2	3	20	**Naringenin**	8.3	1.4	3	20
Nonylphenol	110	95	2	13	**Nonylphenol**	84	84	1	7
Fluoxetine	0.64	0.64	1	7	**Norfluoxetine**	0.51	0.51	1	7
Norfluoxetine	0.77	0.77	1	7	**Sulfamethoxazole**	0.32	0.32	1	7
Risperidone	2.9	2.9	1	7					
Sulfamethoxazole	0.32	0.32	1	7					

Undetected Compounds

There were several compounds selected for evaluation during this project that were not detected in any of the samples analyzed. Table 3.6 summarizes those target compounds that were not detected in discrete samples for each water matrix. Although the pharmaceutical simvastatin was not detected in any water sample, its major metabolite simvastatin hydroxyl acid was detected in source and raw water. The phytoestrogen equol, a metabolite of daidzein, also was not detected in any water samples as well as the potential EDCs, methoxychlor, octachlorostyrene, vinclozolin, and all four BHC isomers. Risperidone was detected at one distribution site and one source water site close to the MRL, but not detected in raw intake water.

Table 3.6
List of compounds not detected

	Source water	Raw intake water	Finished and distribution water
Pharmaceuticals	Simvastatin	Simvastatin	Simvastatin
		Enalapril	Enalapril
			Trimethoprim
			Atorvastatin
			o-Hydroxy atorvastatin
			p-Hydroxy atorvastatin
			Simvastatin hydroxy acid
			Diclofenac
			Naproxen
		Risperidone	
Steroids	Testosterone		Testosterone
	Ethynylestradiol		Ethynylestradiol
			Estrone
			Estradiol
Phytoestrogens	Equol	Equol	Equol
		Matairesinol	Matairesinol
			Formononetin
			Biochanin A
			Apigenin
			Coumestrol
			Chrysin
Potential EDCs	α-BHC	α-BHC	α-BHC
	β-BHC	β-BHC	β-BHC
	γ-BHC	γ-BHC	γ-BHC
	δ-BHC	δ-BHC	δ-BHC
	Vinclozolin	Vinclozolin	Vinclozolin
	Octachlorostyrene	Octachlorostyrene	Octachlorostyrene
	Methoxychlor	Methoxychlor	Methoxychlor
		BHA	BHA
		Diazinon	Diazinon
		Traseolide	Traseolide
			octylphenol
			Benzophenone
			Musk Ketone
			Tonalide
			Bis(2-ethylhexyl) phthalate
			Butylbenzyl phthalate

Occurrence in Wastewater Samples

The complete results of target compounds in wastewater are presented in Appendix B. WWTP4 is a water reuse facility that receives secondary wastewater; therefore, influent in this case has already undergone secondary treatment which is responsible for the lower detection of target compounds. Table 3.7 lists several representative compounds with either high or low removals through the WWTPs. Persistent compounds include carbamazepine, phenytoin, meprobamate, and TCEP. Their persistence and/or stability have been demonstrated by many studies during both drinking water and wastewater treatment processes (Clara, Strenn, and Kreuzinger 2003, Westerhoff et al. 2005, Castiglioni et al. 2006). Conversely, naproxen and trimethoprim exhibited higher removal efficiencies, which explain the relatively low occurrence of these two compounds in water samples.

Table 3.7
Occurrence of selected pharmaceuticals and potential EDCs in WWTPs (ng/L)

	WWTP1		WWTP2		WWTP3		WWTP4		
	Influent	MBR permeate	Influent	Secondary effluent	Influent	Tertiary effluent	Influent	Train 1 effluent	Train 2 effluent
Carbamazepine	110	140	99	210	160	220	280	220	<0.50
Phenytoin	54	140	94	150	200	250	250	220	<1.0
Meprobamate	230	260	740	330	780	330	310	330	0.39
Naproxen	6800	140	13000	13	12000	<0.50	230	120	<0.50
TCEP	51	160	450	370	1100	940	600	480	<50
Trimethoprim	250	<10	320	35	840	2.8	610	190	1.7

Field and Laboratory Blanks

During each sampling event, field (travel) blanks were collected along with actual samples. The field blanks follow the entire sampling and collection procedure that the water samples endure. Blanks are conducted to determine if samples may have become contaminated during the collection and transport process. It is important to note that initial reporting limits are calculated at the beginning of method development and based on instrument sensitivity. These calculated reporting limits do not take into account the sample collection process or unforeseen contamination. Throughout the project it was observed that a few compounds were detected in travel blanks at levels just above the initial reporting limits. These compounds included risperidone, nonylphenol, and bis(2-ethylhexyl) phthalate. Because of these observations, a statistical test was performed on all travel and laboratory blanks collected throughout the study (n=68). After this evaluation, revised reporting limits were set at 2 standard deviations above the mean concentration for these three compounds (Table 3.8). All results were adjusted to reflect the final reporting limits. Table 3.9 provides the compounds that were detected in the blanks with corresponding maximum values, median values, and frequency of detection. A summary of laboratory blanks is shown in Table 3.10. These results demonstrate that contamination rarely occurred and only at levels near the MRLs. Additional QA/QC information are provided in Appendix A.

Table 3.8
Reporting limit comparison for compounds detected in travel and laboratory blanks (n=68)

Description	Initial RL (ng/L)	Final RL (ng/L)
Risperidone	0.25	2.5
Nonylphenol	50	80
Bis(2-ethylhexyl) phthalate	50	120

Table 3.9
Results for field blanks (n=28)

Description	Max	Median	Frequency	
	ng/L	ng/L	#	%
Butylbenzyl phthalate	59	55	2	7.1
Sulfamethoxazole	0.41	0.41	1	3.6
Fluoxetine	0.68	0.68	1	3.6
Norfluoxetine	0.63	0.63	1	3.6
Atrazine	0.29	0.29	1	3.6
Bisphenol A	5.5	5.5	1	3.6
Triclosan	1.0	1.0	1	3.6
Progesterone	0.68	0.68	1	3.6
Daidzein	1.4	1.4	1	3.6
Naringenin	1.6	1.6	1	3.6
BHT	44	44	1	3.6

Table 3.10
Results for laboratory blanks (n=40, *n=36)

Compound	Max	Median	Frequency	
	ng/L	ng/L	#	%
Atenolol	1.1	0.85	2	5.0
Triclosan	3.2	2.1	2	5.0
Estrone	0.35	0.35	1	2.5
Genistein*	1.4	1.4	1	2.5
Daidzein*	1.1	1.1	1	2.5
Naringenin*	2.5	2.5	1	2.5
Bis(2-ethylhexyl) phthalate	130	130	1	2.5

CHAPTER 4
OCCURRENCE OF TARGET COMPOUNDS IN FOOD

SELECTION OF FOOD ITEMS

In order to provide context for the estradiol equivalent units (EEq) values determined in water samples, a series of food/beverage items also were evaluated for the suite of 62 target analytes. The food/beverage items were included in this study only to provide a reference point for perspective regarding the relevance of in vitro bioassay data in public health protection. Food and beverage items were purchased from local vendors in the Southern Nevada area. A limited number of items were chosen as evaluation of food and beverage items are not to be interpreted as a final comprehensive survey, but rather should be viewed as a screening of selected items from one demographic. Food and beverage brands were based on availability and popularity of brand name, whereas more obscure brands were avoided. A minimum of two different brands were evaluated for each type of food/beverage item, so comparisons between brands was not investigated. Selected items included bottled water, apple juice, vegetable juice, green tea, coffee, beer, milk, soy milk, baby formula, soy baby formula, and soy sauce.

METHOD SUMMARY

Storage and Preparation

Perishable items were stored at 4 °C until extraction, with all items remaining in their original containers. Samples were extracted within one month of purchase and prior to the expiration date. All food and beverage items were purchased prepared and ready for consumption except for green tea, which was prepared according to manufacture's directions. Extraction volumes were based upon suggested serving sizes.

Extractions

Automated SPE was used to extract target compounds from food and beverage items when applicable, using similar SPE methods as employed for water samples. Some extractions required additional steps for sample preparation. When SPE was not applicable, such as the case for soy sauce, apple juice, and vegetable juice, liquid/liquid extractions were performed using dichloromethane (DCM). Highlighted here are the main differences between the extraction methods, with a summarized comparison of all extractions shown in Figure 4.1. More detailed information on extractions methods can be found in Appendix A. Isotopically-labeled compounds used for surrogate and isotope dilution standards were spiked prior to sample extraction in each food item.

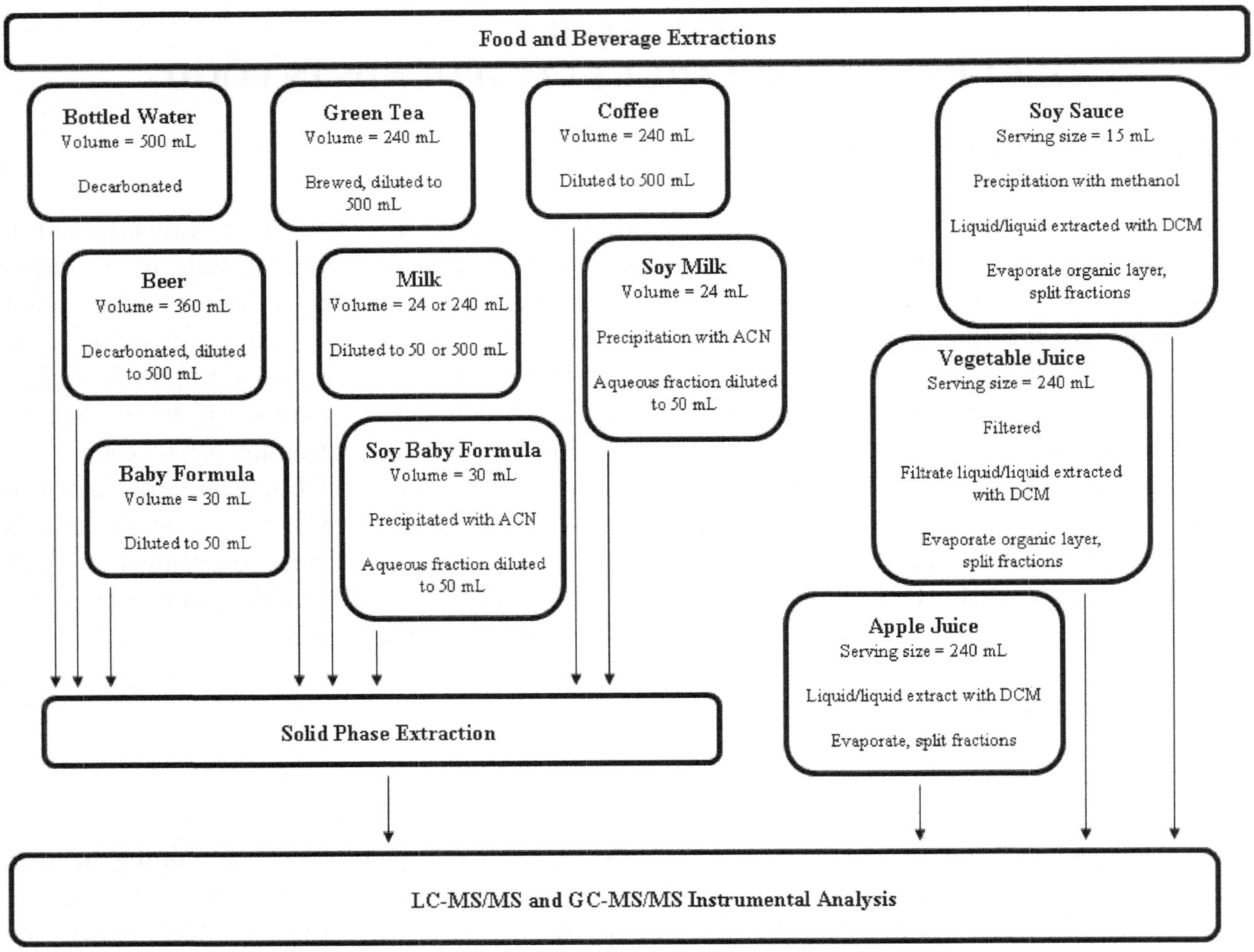

Figure 4.1 Summary of food and beverage extraction methods

Bottled Water – A volume of 500 mL was extracted by SPE. When necessary, carbonation was removed prior to extraction by continually shaking and venting the sample.

Green Tea – Green tea samples were brewed according to manufacturer directions. A recommended serving size of 240 mL was diluted to 500 mL with reagent water and extracted by SPE.

Coffee – Brewed coffee was purchased directly from vendors. A recommended serving size of 240 mL was diluted to 500 mL with reagent water and extracted by SPE. Because of instrumental complications with the extracted coffee matrix, only E-screen analysis was performed.

Beer – Prior to extraction, carbonation was removed from samples by shaking and venting to the atmosphere. A recommended serving size of 360 mL was diluted to 500 mL with reagent water and extracted by SPE.

Milk – A recommended serving size of 240 mL was used for pharmaceuticals and potential EDCs, which was diluted to 500 mL with reagent water and extracted by SPE. Due to matrix interference and expected high concentrations, only a 24 mL aliquot was used for steroids/phytoestrogens extractions, which was diluted to 50 mL and extracted by SPE.

Soy Milk – Although the recommended serving size is 240 mL, only a 24 mL aliquot was used for extractions. To minimize matrix interferences, precipitation was performed by adding 26 mL acetonitrile (ACN) to each sample and centrifuged. The supernatant was decanted and concentrated with nitrogen gas. The remaining aqueous fraction was diluted to 50 mL with reagent water and extracted by SPE.

Milk Based Infant Formula – Since serving sizes of milk based infant formula (ready to use) vary depending on age and weight of the baby, extractions were performed based on manufacturer's directions for a 1 ounce (30 mL) preparation and nutritional guideline. An aliquot of 30 mL was diluted to 50 mL and extracted by SPE.

Soy Based Infant Formula – Similar to milk based infant formula, a 30 mL aliquot of soy based infant formula was used for extractions. In order to minimize matrix interferences, precipitation was performed by adding 30 mL ACN to each sample and centrifuged. The supernatant was decanted and concentrated with nitrogen gas. The remaining aqueous fraction was diluted to 50 mL with reagent water and extracted by SPE.

Soy Sauce – A 15 mL serving size was added to 100 mL of methanol and chilled. Samples were centrifuged and the supernatant was decanted and concentrated with nitrogen gas. The remaining water fraction was liquid/liquid extracted with DCM. The organic layer was collected and concentrated. Extracts were split into two equal fractions and solvent exchanged into solvents more suitable for instrumental analysis.

Vegetable Juice – A recommended serving size of 240 mL was first centrifuged. The supernatant was decanted and filtered. The filtrate was then liquid/liquid extracted with DCM. The collected organic extract was concentrated and split into two equal fractions. Each fraction was solvent exchanged into a solvent more suitable for instrumental analysis.

Apple Juice – A recommended serving size of 240 mL was liquid/liquid extracted with DCM. The collected organic extract was concentrated with nitrogen gas and split into two equal fractions. Each fraction was solvent exchanged into a solvent more suitable for instrumental analysis.

OCCURRENCE IN FOOD ITEMS

Results

Tables 4.1 through 4.9 provide food/beverage testing results and associated QA/QC. With the exception of phytoestrogens, only results for detected compounds are shown. When duplicate or triplicate samples were extracted, the average is reported. A non-detect value in a replicate was averaged by using one-half the reporting limit value. Reporting limits are adjusted according to the actual sample volume used for extraction (based on serving size) and therefore MRLs vary with each food item. Because phytoestrogens and steroid hormones were expected to dominate detection in the chosen food items, extractions were specifically tailored for these compounds. Therefore, some recoveries for the other classes of compounds were not ideal.

Table 4.1
Occurrence of target compounds detected in soy sauce (ng/L)

Description	Brand 1	Brand 2	Brand 3	Brand 4	Matrix Spike	DI Spike Recovery	DI Blank
Phytoestrogens							
Apigenin	<330	2000	<330	2100	60%	67%	<330
Biochanin A	<330	<330	<330	<330	92%	67%	<330
Chrysin	<330	<330	<330	<330	88%	66%	<330
Coumestrol	<330	390	540	<330	80%	56%	<330
Daidzein	4500000	400000	2500000	7700000	88%	44%	<330
Equol	<3300	<3300	<3300	<3300	34%	68%	<3300
Formononetin	<330	810	6800	920	101%	67%	<330
Genistein	3400	1600000	800000	2200000	99%	60%	<330
Glycitein	200000	1100000	580000	1300000	87%	52%	<330
Matairesinol	<1650	<1650	<1650	<1650	14%	69%	<1650
Naringenin	<330	20000	1200	43000	51%	58%	<330
Potential EDCs							
BHT	5500	3300	20000	4500	150%	122%	3500
Benzophenone	1900	12000	<1750	<1750	65%	116%	<1750
Galaxolide	<1750	<1750	<1750	2000	94%	81%	<1750

Table 4.2
Occurrence of target compounds detected in apple juice (ng/L)

Description	Brand 1	Brand 2	Brand 3	Brand 4	Matrix Spike	DI Spike Recovery	DI Blank
Phytoestrogens							
Apigenin	260	940	<210	420	99%	52%	<210
Biochanin A	<210	<210	<210	<210	113%	69%	<210
Chrysin	260	580	<210	410	116%	72%	<210
Coumestrol	<210	<210	<210	<210	107%	64%	<210
Daidzein	<210	<210	<210	<210	46%	26%	<210
Equol	<2100	<2100	<2100	<2100	110%	53%	<2100
Formononetin	<210	<210	<210	<210	116%	69%	<210
Genistein	<210	<210	<210	<210	104%	53%	<210
Glycitein	<210	<210	<210	<210	117%	64%	<210
Matairesinol	<1100	<1100	<1100	<1100	112%	67%	<1100
Naringenin	2600	8800	430	7400	120%	49%	<210
Potential EDCs							
Butylbenzyl-phthalate	<250	<250	<250	350	96%	124%	<250
Nonylphenol	10000	26000	16000	26000	N/A*	N/A*	4400
TCPP	<250	510	<250	540	86%	110%	<250

* Spike not observable above analyte background concentrations

Table 4.3
Occurrence of target compounds detected in vegetable juice (ng/L)

Description	Brand 1	Brand 2	Brand 3	Brand 4	Matrix Spike	DI Spike Recovery	DI Blank
Phytoestrogens							
Apigenin	210	340	630	1700	16%	29%	<210
Biochanin A	<210	<210	<210	<210	8%	99%	<210
Chrysin	<210	<210	<210	<210	9%	99%	<210
Coumestrol	<210	<210	<210	<210	9%	48%	<210
Daidzein	<210	<210	<210	<210	9%	10%	<210
Equol	<2100	<2100	<2100	<2100	23%	80%	<2100
Formononetin	<210	<210	<210	<210	16%	89%	<210
Genistein	<210	<210	<210	<210	12%	36%	<210
Glycitein	<210	<210	<210	<210	29%	53%	<210
Matairesinol	<1100	<1100	<1100	<1100	52%	72%	<1100
Naringenin	910000	540000	880000	580000	N/A*	29%	<210
Steroids							
Testosterone	<5.0	15	<5.0	<5.0	14%	38%	<5.0
Potential EDCs							
Linuron	<25	<25	100	<25	141%	104%	<20
Nonylphenol	700	1000	1100	1400	50%	45%	640

* Spike not observable above analyte background concentrations

Table 4.4
Occurrence of target compounds detected in green tea (ng/L)

Description	Brand 1	Brand 2	Brand 3	Brand 4	Matrix Spike	DI Spike Recovery	DI Blank
Phytoestrogens							
Apigenin	2900	2700	9400	5600	N/A*	72%	<1.0
Biochanin A	<21	<21	<21	24	75%	89%	<1.0
Chrysin	<21	<21	44	56	69%	83%	<1.0
Coumestrol	<21	<21	<21	520	87%	93%	<1.0
Daidzein	<21	<21	<21	51	37%	85%	<1.0
Equol	<210	<210	<210	<210	36%	95%	<10
Formononetin	250	<21	<21	390	71%	103%	<1.0
Genistein	<21	<21	26	130	65%	83%	<1.0
Glycitein	100	160	65	920	42%	90%	<1.0
Matairesinol	5000	6900	1100	3300	12%	103%	<5.0
Naringenin	17000	12000	12000	4100	N/A*	98%	1.1
Potential EDCs							
BHA	<53	<53	98	<53	9%	60%	<25
Nonylphenol	960	690	1700	650	50%	82%	<80

* Spike not observable above analyte background concentrations

Table 4.5
Occurrence of target compounds detected in beer (ng/L)

Description	Brand 1	Brand 2	Brand 3	Brand 4	Brand 5	Matrix Spike	DI Spike Recovery	DI Blank
Phytoestrogens								
Apigenin	5200	5150	4200	3000	4800	103%	71%	<10
Biochanin A	<150	<150	<150	<150	<150	90%	78%	<10
Chrysin	<150	<150	<150	<150	<150	104%	78%	<10
Coumestrol	<150	<150	<150	<150	<150	114%	80%	<10
Daidzein	<150	<150	180	<150	360	102%	103%	<10
Equol	<1500	<1500	<1500	<1500	<1500	97%	113%	<100
Formononetin	<150	<150	<150	<150	<150	114%	99%	<10
Genistein	<150	<150	300	<150	540	111%	94%	<10
Glycitein	<150	<150	<150	<150	<150	109%	92%	<10
Matairesinol	<750	<750	<750	<750	<750	129%	96%	<50
Naringenin	9600	980	1700	400	1000	106%	107%	<10
Steroids								
Estrone	<500	<500	<500	940	<500	42%	94%	<50
Potential EDCs								
Nonylphenol	2200	2450	1900	1900	1500	42%	58%	<500

Table 4.6
Occurrence of target compounds detected in milk (ng/L)

Description	Brand 1	Brand 2	Brand 3	Matrix Spike	DI Spike Recovery	DI Blank
Pharmaceuticals						
Atenolol	64	57	47	115%	107%	<0.25
Steroids						
Testosterone	<25	33	26	75%	93%	<10
Progesterone	2700	3700	3100	57%	72%	<10
Phytoestrogens						
Apigenin	68	30	70	87%	69%	<10
Biochanin A	<21	<21	<21	82%	77%	<10
Chrysin	<21	<21	<21	83%	80%	<10
Coumestrol	190	100	180	88%	74%	<10
Daidzein	<21	24	<21	104%	99%	<10
Equol	<21	<21	<21	109%	96%	<100
Formononetin	<21	<21	<21	102%	97%	<10
Genistein	35	35	21	97%	90%	<10
Glycitein	<21	<21	<21	93%	94%	<10
Matairesinol	<21	<21	<21	96%	98%	<50
Naringenin	77	32	41	106%	98%	<10

Table 4.7
Occurrence of target compounds detected in soy milk (ng/L)

Description	Brand 1	Brand 2	Brand 3	Matrix Spike	DI Spike Recovery	DI Blank
Phytoestrogens						
Apigenin	2300	7900	<2100	79%	70%	<210
Biochanin A	<2100	<2100	<2100	74%	75%	<210
Chrysin	<2100	<2100	<2100	75%	77%	<210
Coumestrol	<2100	<2100	<2100	77%	62%	<210
Daidzein	2100000	2700000	1100000	117%	97%	<210
Equol	<21000	<21000	<21000	90%	91%	<2100
Formononetin	<2100	4000	<2100	91%	91%	<210
Genistein	2200000	3300000	1100000	140%	94%	<210
Glycitein	110000	240000	71000	65%	83%	<210
Matairesinol	<11000	<11000	<11000	86%	89%	<1100
Naringenin	69000	140000	47000	84%	92%	<210
Potential EDCs						
BHA	4700	<625	<625	52%	47%	<625
BHT	18000	15000	15000	41%	71%	1200
Benzophenone	910	<625	<625	98%	106%	<625
Nonylphenol	3100	<625	<625	34%	32%	<1250

Table 4.8
Occurrence of target compounds detected in baby formula (ng/L)

Description	Brand 1	Brand 2	Matrix Spike	DI Spike Recovery	DI Blank
Steroids					
Progesterone	170	120	30%	81%	<20
Phytoestrogens					
Apigenin	34	48	93%	68%	<10
Biochanin A	<20	<20	74%	73%	<10
Chrysin	<20	<20	74%	72%	<10
Coumestrol	35	38	96%	74%	<10
Daidzein	2600	2300	101%	88%	<10
Equol	<200	<200	104%	107%	<100
Formononetin	24	28	108%	91%	<10
Genistein	1500	3400	104%	86%	<10
Glycitein	230	280	91%	84%	<10
Matairesinol	<100	1100	99%	92%	<50
Naringenin	590	76	113%	89%	<10
Potential EDCs					
Bisphenol A	<1000	3100	145%	104%	<100

Table 4.9
Occurrence of target compounds detected in soy baby formula (ng/L)

Description	Brand 1	Brand 2	Matrix Spike	DI Spike Recovery	DI Blank
Phytoestrogens					
Apigenin	2200	4500	90%	69%	<10
Biochanin A	170	740	91%	77%	<10
Chrysin	28	53	83%	76%	<10
Coumestrol	590	1400	100%	83%	<10
Daidzein	480000	950000	138%	96%	14
Equol	<200	<200	101%	99%	<100
Formononetin	1500	3100	110%	90%	<10
Genistein	630000	1100000	121%	89%	31
Glycitein	100000	130000	116%	90%	<10
Matairesinol	230	1200	114%	95%	<50
Naringenin	40000	47000	99%	100%	<10
Potential EDCs					
Bisphenol A	5500	<1000	126%	107%	<100
Nonylphenol	11000	<10000	38%	5%	<1000

As the results tables show, many of these items did have detectable levels of phytoestrogens, particularly items containing soy. Soy sauce was observed to have many phytoestrogens present in mg/L (ppm) ranges, including daidzein, genistein, and glycitein. Similarly, soy milk and soy baby formula also had various phytoestrogens found in the mg/L range. Phytoestrogens were detected in all the food and beverage items tested, with the exception of bottled water. Bottled water resulted in non-detect for all target analytes and therefore data is not shown.

A significant portion of the total phytoestrogens in food/beverages evaluated was due to only a few compounds. Figure 4.2 displays the primary phytoestrogens detected in food and beverage items. Although levels of phytoestrogens in cow's milk were relatively low, progesterone was found at concentrations of 2.7 – 3.7 μg/L. Also common in many food items were BHA and BHT, which are commonly used as a food preservatives. It is important to point out that these food extractions underwent more complex sample preparation and extraction techniques than the typical utility water extractions, and may have encountered more possible sources for contamination. Another observation to note is the increased complexity of the matrices related to the food items during instrumental analysis resulting in a higher probability of interferences, false positives and carryover. Therefore, a few compounds detected in food and beverage items were also detected in the DI reagent water blanks, and results should be interpreted accordingly.

It is also important to keep in mind that the objective of the food analyses was not to suggest that these items, especially soy based products with high concentations of phytoestrogens, are dangerous or of concern for human health and consumption. Instead, the goal was to put in perspective the trace amounts and low occurrence of these compounds found in drinking water versus other sources that are deemed natural and considered relatively safe. With such a limited amount of data, no conclusions should be made from this study about human risks from any detected compounds in food and beverage items tested.

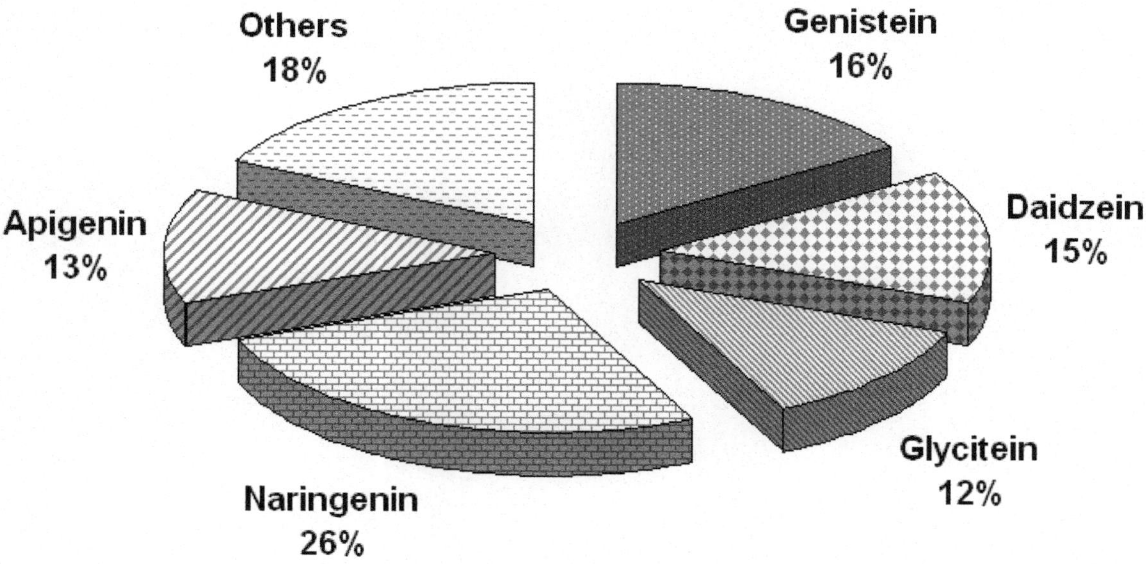

Figure 4.2 Distribution of total phytoestrogens detected in food and beverage items

CHAPTER 5
OCCURRENCE OF IN VITRO ESTROGENICITY

BACKGROUND

InVitro Bioassay

The mechanism of action of steroid hormones is fairly well understood. Steroid hormones such as estradiol exert their activity by binding to specific receptors which translocate into the nucleus and increase or decrease gene transcription. Because this mechanism is established, in vitro bioassays have been developed that can indicate whether compounds are binding to the receptor. In the MCF-7 breast cancer cell line, a response to compounds binding to the estrogen receptor is an increase in cell proliferation, i.e. a marked increase in the number of cells. The E-screen is an in vitro bioassay developed to identify compounds that bind to the estrogen receptors that uses the proliferation of MCF-7 cells as the endpoint indicating estrogen activity (Soto et al. 1995). The E-screen has been used to identify many estrogenic compounds including surfactants, plasticizers, and pesticides (Sonnenschein and Soto 1998). The MCF-7 cell line also has been used to determine whether environmental samples, such as surface waters and wastewaters contain compounds that are estrogenic (Snyder, Snyder et al. 2000; Snyder, Villeneuve et al. 2001a; Hemming, Barman et al. 2003; Drewes, Hemming et al. 2005; Schappell 2006). Although the E-screen can not indicate which specific compounds are responsible for the activity, it has the benefit of indicating the biological activity of a sample even if the compounds of interest are as yet unidentified.

Method Summary

MCF-7 Cell Culturing

A breast cancer cell line (MCF-7) was continuously cultured for use in the E-Screen. The cells were obtained from Drs. Sonnenschein and Soto at Tufts University (Boston, Mass.). The MCF-7 cells were cultured in Dulbecco's modified Eagle's Medium (MP Biomedicals, Solon, Ohio) with 5% fetal bovine serum (FBS) (Hyclone Laboratories, Logan, Utah) at 37°C and 6.5% CO_2 in 75 cm^2 tissue culture flasks. Media was changed every 2-3 days and the cells were subcultured every seven days. Stocks of MCF-7 cells were stored in liquid nitrogen and new cells were thawed to replace cells that had undergone approximately 25 passages.

E-Screen

For the E-Screen, seven days following sub-culturing, cells from a 75-cm^2 flask were trypsinized and counted on an EPICS XL flow cytometer (Coulter Corporation, Miami, Fla.) Cell counts were achieved using FlowCount beads (Coulter Corporation, Miami, Fla.) an FDA-approved microbead standard originally developed for enumeration of blood cells. Once the concentration of the cells was determined, cells were then seeded in 24-well plates to achieve 20,000-30,000 cells per well, in 1-mL of culturing media. Twenty-four hours after seeding, the media was removed and experimental media was added.

The experimental media was Dulbecco's Modification of Eagles Medium without phenol red (Mediatech Inc., Herndon, Va.). The 5% FBS used in this media was stripped of steroids with a charcoal dextran (CD) stripping procedure described in Payne et al. (Payne et al. 2000). Briefly, FBS was incubated with activated charcoal and dextran followed by centrifugation and filtration (0.2μm) to remove the charcoal. This experimental media is referred to as CD-media.

For plates containing the standards, the MCF-7 cells were exposed to 15 concentrations of 17β-estradiol (Sigma Chemical Co. St. Louis Mo.) in CD-media ranging from 0.14 to 2724 ng/L. All concentrations were applied to four wells, with three plates needed to obtain a complete standard curve. Four control wells were included on each plate. Samples were added to CD-media at no greater than 1% (v/v) and a dilution series was made if necessary to obtain values that would fall within the standard curve. Each concentration was applied to four wells at a volume of 0.5 mLs. In one of those wells, estradiol was added to obtain a concentration of 27.2 ng/L in the well. This positive control was used to determine whether anything in the sample was preventing the growth of the cells. Interference in this positive control can occur due to cytotoxicity of the sample or could also occur due to the presence of an estrogen antagonist. Again, each plate contained 4 blank control wells.

In order to determine the relative potency of other estrogenic chemicals in the E-screen, concentration-response curves for compounds of interest were made using at least eight different concentrations. Each concentration was applied to the plate in four wells and four control wells were included on each plate.

After five days of incubation, the media was removed and the cells were fixed with 10% trichloroacetic acid (TCA) solution. The wells were rinsed with water and again allowed to dry. Cell proliferation was measured by the sulphorhodamine B (SRB) (Sigma Chemical Co., St. Louis, Mo.) protein assay, which estimates total cell numbers by measuring total protein content. SRB solution (0.4% SRB dye in 1% acetic acid/99% distilled water) was added to each well to stain the cells. The residual SRB dye was removed by rinsing with 1% acetic acid after 20-30 minutes. The remaining dye was redissolved using 10mM Tris solution (pH 10.5) and read at a wavelength of 515 nm.

To determine the estradiol equivalents (EEq) of samples, the standard curve was fit with a 4-parameter logistic equation. This is done with the Softmax PRO v. 2.6 software package, the software analysis package for the Molecular Devices microplate reader (Sunnyvale Calif.). The estradiol equivalents for unknown samples were calculated from this equation. The estradiol equivalent was corrected for the dilution of the sample for use in the E-screen (at least 100-fold), and the concentration of the water sample during the extraction procedure. Dilutions resulting with values closest to the concentration which caused 50% of the maximum proliferation (EC_{50}) for the standard curve were reported and standard deviations were calculated from the three replicate wells in an individual assay. The EC_{50} for the concentration-response curves of the other estrogenic compounds was calculated using Microcal Origin v. 4.1 (Northampton, Mass.). The potency for each chemical was determined relative to the EC_{50} of 17β-estradiol. Figure 5.1 compares the E-screen response for 17β-estradiol and several selected phytoestrogens. Figure 5.2 compares the E-screen response for 17β-estradiol and several selected potential EDCs and compounds with estrogenicity.

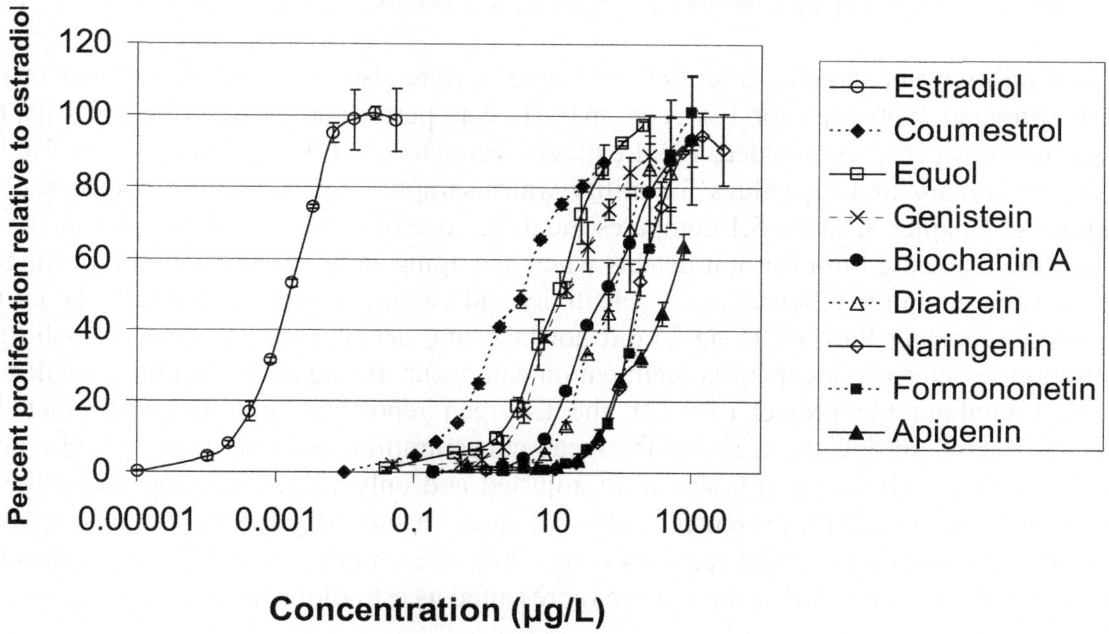

Figure 5.1 E-screen concentration response curves for 17β-estradiol and several phytoestrogens.

For curves with one run, standard deviations represent the responses of triplicate wells. For curves with two runs, standard deviations represent the responses of the two separate runs. Concentrations indicating interference in the positive control were not included in the curve.

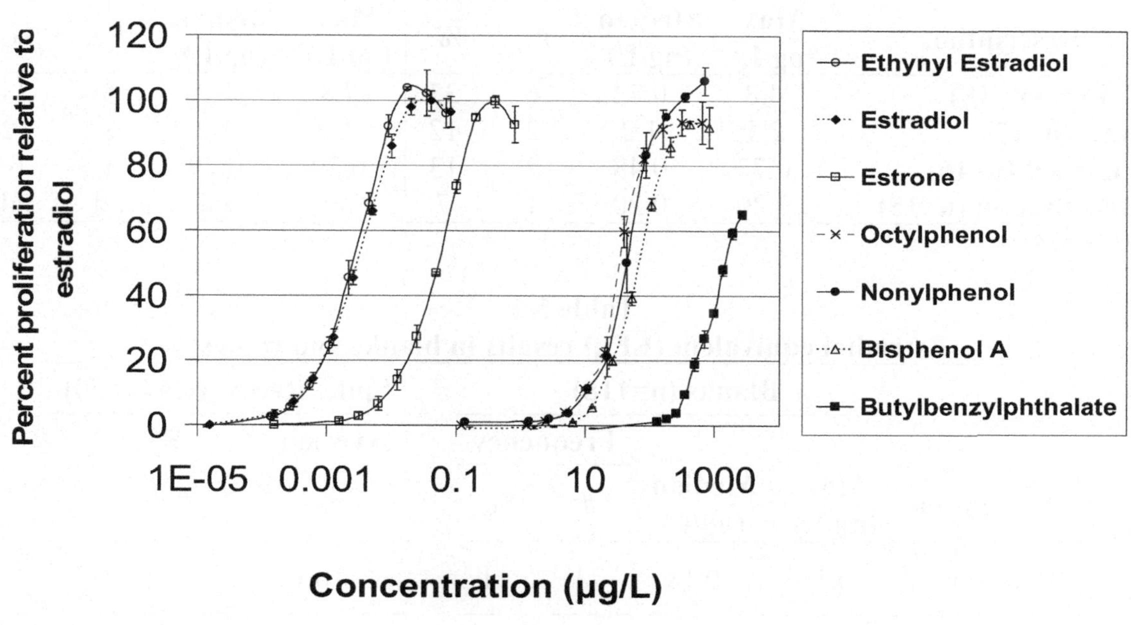

Figure 5.2 E-screen concentration response curves for 17β-estradiol and several other estrogenic compounds.

Values represent the mean ± standard deviation of two separate runs. Concentrations indicating interference in the positive control were not included in the curve.

OCCURRENCE IN UTILITY WATER - EXTRACTIONS FOR E-SCREEN ANALYSIS

All utility water samples described in Chapter 3 were also extracted for E-screen analysis. Samples extracted (500 mL) for E-screen analysis was performed as described in Chapter 3; however, no surrogates were added. Final extracts were stored at 4 °C until analysis. Table 5.1 provides a summary of EEq values for utility water samples. Similar to the results for target compounds in Chapter 3, Table 5.1 compares the difference of using only detected values versus using half the reporting limit for non-detects when averaging samples. The initial reporting limit for EEq was based on the E-screen assay sensitivity and volume of water extracted. The initial E-screen assay reporting limit (0.03 ng/L) did not take into account travel blanks or solid phase extraction, in which trace levels of contamination can occur. Based on the findings of all blanks analyzed throughout the project (n=118), the E-screen reporting limit was re-evaluated and adjusted to 2 standard deviations above the mean concentration, resulting in a revised reporting limit of 0.16 ng/L. All data and tables were adjusted and only values detected above the final reporting limit are henceforth presented. Table 5.2 shows a summary for the E-screen results for DI reagent water blanks and spike recovery tests. Only one sample out of 118 blanks showed an EEq value greater than the MRL. Results are represented as estradiol equivalents (EEq) in ng/L.

Table 5.1
Estradiol equivalent (EEq) results in utility water

	Non-detected results ignored				Non-detected results averaged using ½ RL			
			Frequency				Frequency	
Description	Max (ng/L)	Median (ng/L)	#	%	Max (ng/L)	Median (ng/L)	#	%
Source (n=18)	4.8	0.72	6	33	4.8	1.2	5	28
Raw (n=17)	2.1	1.2	2	12	2.1	1.2	2	12
Finished (n=16)	0.77	0.48	2	13	0.29	0.29	1	6
Distribution (n=15)	0.20	0.20	1	7	nd	nd	nd	nd

* nd: non-detect

Table 5.2
Estradiol equivalent (EEq) results in blanks and spikes

	Blanks (n=118)				Spike Recoveries (n=50)	
			Frequency		Average	RSD
Description	Max (ng/L)	Median (ng/L)	#	%	%	%
DI water	0.18	0.18	1	1	106	13

OCCURRENCE IN FOOD ITEMS - EXTRACTIONS FOR E-SCREEN ANALYSIS

All food and beverage items described in Chapter 4 also were extracted for E-screen analysis and analyzed as described; however, no surrogate standards were added. Final extracts were kept at 4 °C until analysis. Table 5.3 shows the E-screen results for food and beverage items along with corresponding QA/QC samples. Results also are presented as EEq units in ng/L.

Results for matrix spikes in soy based food and beverage were not observable due to high levels of phytoestrogens.

Table 5.3
Estradiol equivalent (EEq) results in food and beverage products

Description	Brand	EEq (ng/L)	Matrix Spike	DI Blank (ng/L)	DI Spike Recovery
Bottled water	1	<0.16	N/A	<0.16 (n=2)	120% (n=3)
	2	<0.16			
	3	<0.16			
	4	<0.16			
	5	<0.16			
Soy sauce	1	300	94% (n=2)	<4.0 (n=2)	96% (n=1)
	2	140			
	3	28			
	4	510			
Green tea[1]	1	1.5 – 2.1	120% (n=1)	<0.16 (n=2)	100% (n=2)
	2	2.1 – 3.3			
	3	2.5 – 3.8			
	4	2.9 – 4.4			
Apple juice	1	<2.5	94% (n=2)	<0.25 (n=2)	97% (n=1)
	2	0.74			
	3	0.73			
	4	0.79			
Vegetable juice	1	1.9	18% (n=2)	<0.25 (n=1)	99% (n=2)
	2	2.9			
	3	3.3			
	4	2.2			
Milk	1	0.48	64% (n=3)	<0.16 (n=1)	92% (n=1)
		0.81			
	2	0.55			
	3	0.54			
Soy milk	1	1900	N/A*	<0.16 (n=2)	92% (n=2)
	2	4000			
	3	4200			
Baby formula	1	0.73	82% (n=2)	<0.3 (n=2)	65% (n=2)
	2	0.85			
Soy baby formula	1	1900	N/A*	<0.3 (n=2)	100% (n=2)
	2	1500			

(continued)

Table 5.3 (Continued)

Description	Brand	EEq (ng/L)	Matrix Spike	DI Blank (ng/L)	DI Spike Recovery
Coffee	1	12	69% (n=2)	<0.16 (n=2)	110% (n=2)
	2	12			
	3	11			
	4	17			
Beer	1	6.0	N/A*	<0.16 (n=2)	110% (n=2)
	2	2.0			
	3	140			
	4	0.80			
	5	11			

* Spike not observable above analytes background concentrations

[1] EEq values for green tea are expressed as a range due to sample cytotoxicity and subsequent dilutions

CHAPTER 6
RISK ANALYSIS FOR PHARMACEUTICALS

TOXICITY ASSESSMENT OF PHARMACEUTICALS

The toxicity assessment for pharmaceuticals relied on a combination of published toxicity studies, pharmacological information presented in FDA files and package inserts, and information on drug side effect. For example, prior to marketing a pharmaceutical in the U.S., the manufacturer is required to collect data to establish its safety. Studies are conducted to assess the drug's behavior, toxicological effects in animals, and safety and efficacy in humans.

Additionally, published "side effect" profiles provide evidence that unintended effects are known to occur at therapeutic doses. In some cases, studies conducted after marketing a drug reveal unexpected effects over the course of use (e.g., associations with developmental or neurological effects, or cancer development) — the FDA monitors such responses through its Adverse Event Reporting System (AERS) (FDA 2007). The most commonly affected systems are the nervous, blood, cardiovascular, and respiratory systems (Kelly 2001).

DATA TYPES

Several toxicity data types were considered. A typical suite of animal toxicology studies conducted prior to drug marketing includes acute studies (exposure of one day or less to evaluate effects that happen very rapidly after an exposure), subchronic studies (up to three months in duration), chronic studies (repeated longer-term exposure, usually up to two years, to assess the potential for cancer, organ problems, nervous system impairment, reproductive effects, etc.), studies specific to effects on reproduction and offspring development (with administration just prior to, during, and immediately after pregnancy), and genotoxicity studies (to assess whether the agent causes genetic mutations). These studies generally use therapeutic or higher doses that are delivered for a relatively short period of time.

Human clinical trials assessing safety and efficacy at therapeutic doses are also conducted if the results of animal testing suggest a promising efficacy and toxicity profile. In some cases, epidemiological studies have been conducted post-marketing to evaluate certain endpoints that may have become of concern over the course of use (*e.g.*, alleged developmental or neurological effects, or associations with cancer development). Case reports of adverse effects are also often available.

Other information provided may include interaction warnings and notations regarding sensitive populations based on mode of action theory or practice. Drugs are often not studied in children; as such, information about potential risks of exposure to this group may be limited to evidence from animal studies. Low concentration studies are largely limited to studies of aquatic organisms, often involving mixtures (Brooks et al. 2003, Foran et al. 2004, Richards et al. 2004, Robinson, Belden, and Lydy 2005). Continual, life-long exposure to trace levels is a largely unexplored domain of toxicology.

DERIVATION OF ACCEPTABLE DAILY INTAKES (ADIS) FOR NONCANCER EFFECTS

Although no specific regulatory guidance exists prescribing how to assess human health risks of exposure to PPCPs or EDCs in drinking water, standard methods exist for determining exposure levels to environmental contaminants that are not likely to be associated with adverse health effects (WHO 1994, USEPA 2000b, ATSDR 2007). These exposure levels, often termed acceptable daily intakes (ADIs), are typically defined by reviewing animal toxicology data and, for effects other than carcinogenicity, identifying a point of departure upon which to base the ADI. This is typically the highest dose at which an effect is not seen (the "no observed adverse effect level" or NOAEL) or the lowest dose at which an effect is seen (the "lowest observed adverse effect level" or LOAEL). Below this dose, there is no evidence of a statistically or biologically significant increase in adverse effects, although some changes may occur that are not considered adverse (*e.g.*, changes in certain enzyme levels). The point of departure is then divided by uncertainty factors (UFs) to derive an ADI considered protective to broader population groups, including sensitive populations:

$$ADI\,(mg\,/\,kg-d\,) = \frac{NOAEL\ or\ LOAEL\,(\,mg\,/\,kg-d\,)}{UFs}$$

An ADI is commonly defined as the amount of a chemical to which a person can be exposed on a daily basis over an extended period of time (usually a lifetime) without suffering a deleterious effect (USEPA 1993a). ADIs are often presented in terms of dose per day (*e.g.*, mg/d) or dose per unit of body weight per day (*e.g.*, mg/kg-d). Generally, several UFs are applied, individually ranging in value from 3 to 10 with each factor representing a specific area of uncertainty in the available data. For example, if the point of departure is based on an animal study, a factor of 10 may be applied to account for possible differences in responsiveness between animals and humans. A second factor of 10 may account for variation in susceptibility among humans. Other factors may account for database deficiencies, such as when no or minimal information exists on reproductive effects or longer-term exposures. When high-quality toxicity data are available, combined uncertainty factors typically range from 30 to 1,000. Table 6.1 lists the uncertainty factors and their range of values applied in the development of ADIs, as recommended by U.S. EPA (USEPA 2002).

Table 6.1
Standard uncertainty factors and their range of possible values applied in this assessment

Category	Use	Value applied
Interspecies uncertainty/ variability	To extrapolate from an animal study to humans	0 or 10 (a factor of 10 is recommended if the critical study is in animals)
Intraspecies uncertainty/ variability	To account for variations in sensitivity among humans	0, 3, or 10 (if the critical study is based on a known sensitive population group, e.g., developmental data, a factor of 3 is recommended; otherwise, a factor of 10 is recommended)

(continued)

Table 6.1 (Continued)

Category	Use	Value applied
LOAEL to NOAEL	To extrapolate from a LOAEL to a NOAEL.	0 or 10 (a factor of 10 is recommended if the point of departure is a LOAEL)
Study duration	To extrapolate from data obtained in a study with less-than-lifetime exposure to lifetime exposure (i.e., extrapolating from subchronic to chronic exposure)	0, 3, or 10 (a factor of 10 is recommended to extrapolate from a subchronic to chronic study, or 3 if the critical study is a reproductive/ developmental study, and the fetus/children is the most sensitive population group)
Database	To account for limitations in the available quantity or quality of data	0, 3, 10 (a factor of 3 is recommended if either a prenatal study or a 2-generation reproduction study is missing from the database, or 10 if both are missing)

Source: USEPA 2002

Some potential effects of pharmaceuticals are difficult to predict. For example, some pharmaceuticals (*e.g.*, penicillin and sulfa antibiotics, NSAIDs) are known to cause allergic reactions in humans that could range from rashes or hives and to respiratory problems or anaphylaxis, but predicting the likelihood of allergic effects is difficult because people sensitized to a compound can respond to it at extremely low levels of exposure. Additionally, the possibility of microbial resistance induced by antibiotics in the environment is a subject of some controversy. Microbial resistance to antibiotics has been noted in surface water and sewage effluent (Kelch and Lee 1978; Campeau, Gulli et al. 1996; Ash, Mauck et al. 2002); however, (Ayscough et al. 2000) concluded that the most likely cause is not the induction of resistance caused by exposure to low levels of antibiotics, but the excretion of resistant organisms by humans and animals receiving antibiotic treatment.

Toxicity data for each of the target analytes of interest are reviewed in Appendix E. For each, the study that yielded the highest NOAEL and the lowest LOAEL was identified, and then between these, the study that yielded the lowest ADI after application of appropriate uncertainty factors was selected. The final selected studies, the corresponding uncertainty factors, and the resulting ADIs are listed in Table 6.2.

Table 6.2
Toxicological data used to develop acceptable daily intakes (ADIs) and drinking water equivalent levels (DWELs) for noncancer endpoints for target pharmaceuticals

Drug	Study duration	Effect dose (mg/kg-d)	Effect	Reference	UF (animal to human)	UF (intra-species)	UF (LOAEL to NOAEL)	UF (subchronic to chronic)	UF (dbase)	Composite UF	ADI (mg/kg-d)	DWEL (ppb)
Atenolol	**Gestation**	0.80 (LOAEL)	Developmental, human	(Lip, Beevers et al. 1997; Lydakis, Lip et al. 1999; Bayliss, Churchill et al. 2002; Tabacova, Little et al. 2003)	---	3	10	3	3	300	0.0027	93
Atorvastatin	**GD7-PND21**	20 (LOAEL)	Developmental, rat	(Henck et al. 1998)	10	3	10	3	3	3,000	0.0067	230
Carbamazepine	**Gestation**	3.0 (LOAEL)	Developmental, human	(Samren, van Duijn et al. 1997; Samren, van Duijn et al. 1999; Hernandez-Diaz, Werler et al. 2000; Hernandez-Diaz, Werler et al. 2001)	---	3	10	3	3	300	0.010	350
Diazepam	**Gestation**	1.0 (LOAEL)	Developmental, rat	(Silva and Palermo-Neto 1999)	10	3	10	3	---	1,000	0.001	35
Diclofenac	**Gestation**	20 (NOAEL)	Developmental, mouse	(Novartis 2002)	10	3	---	3	3	300	0.067	2,30 0
Enalapril	**Gestation**	0.070 (LOAEL)	Developmental, human	(Tabacova and Kimmel 2001, Tabacova 2005)	---	3	10	3	3	300	0.00023	8.2
Fluoxetine	**Gestation**	0.29 (LOAEL)	Developmental, human	(NTP 2004)	---	3	10	3	3	300	0.00097	34
Gemfibrozil	**GD15-PND21**	92 (LOAEL)	Developmental, rat	(Fitzgerald, Petrere, and de la Iglesia 1987)	10	3	10	3	3	3,000	0.031	1,10 0
Meprobamate	**13 wks**	75 (NOAEL)	Systemic, mouse	(NTP 2000)	10	10	---	10	10	10,000	0.0075	260

(continued)

Table 6.2 (Continued)

Drug	Study duration	Effect dose (mg/kg-d)	Effect	Reference	UF (animal to human)	UF (intra-species)	UF (LOAEL to NOAEL)	UF (subchronic to chronic)	UF (dbase)	Composite UF	ADI (mg/kg-d)	DWEL (ppb)
Naproxen	**Gestation**	170 (NOAEL)	Repro/ Develop, mouse	(Roche 2006)	10	3	---	3	3	300	0.570	20,000
Phenytoin	**PND2-4**	17.5 (NOAEL)	Developmental, mouse	(Ohmori et al. 1997)	10	3	---	3	3	300	0.058	2,000
Risperidone	**Gestation/ Lactation**	0.16 (LOAEL)	Reproductive, rat	(Drugs.com 2006g)	10	3	10	3	3	3,000	0.000053	1.9
Simvastatin	**48 wks**	0.2 (LOAEL)	Developmental, human (children)	(de Jongh et al. 2002)	---	3	10	3	3	300	0.00067	23
Sulfa-methoxazole	**Gestation**	512 (NOAEL)	Developmental, rat	(Pharmaceuticals 2005)	10	3	---	3	10	1,000	0.51	18,000
Triclosan	**2 yr**	75 (NOAEL)	Systemic, hamster	(Barbolt 2002)	10	10	---	---	10	1,000	0.075	2,600
Trimethoprim	**Gestation**	192 (NOAEL)	Developmental, rat	(Pharmaceuticals 2005)	10	3-	---	3	10	1,000	0.19	6,700

DERIVATION OF ACCEPTABLE DAILY INTAKES (ADIS) FOR CANCER EFFECTS

Some pharmaceuticals have been shown to be carcinogenic in high dose animal bioassays conducted as part of the drug development process. Carcinogenicity to humans could be a concern for these chemicals, particularly at low chronic exposure levels. For example, gemfibrozil, a drug taken to reduce blood lipid levels, was shown in a two year cancer bioassay in rats to produce liver tumors, although it is not clear that the required metabolic step that produces reactive chemical metabolites (peroxisome proliferation) is seen in humans (Fitzgerald et al. 1981). For chemicals that show positive evidence of carcinogenicity in high dose animal studies, linear extrapolation models can be used to predict the tumorigenic response at low doses— use of these types of models is recommended as a default for tumor sites where the mode of action is unknown, and assume a linear relationship between risk and dose at low doses (USEPA 2005). The slope of the risk/dose line, known as the slope factor (SF), is an upper-bound estimate of risk per increment of dose (e.g., per 1 mg/kg-day of exposure) that can be used to estimate risk probabilities for different exposure levels.

In this assessment, if sufficient data on tumor incidence per dose level were available for a given compound with evidence of carcinogenicity in animal bioassays, a standard one-hit model was used to estimate a SF. The one hit model is based on the mechanistic argument that a carcinogenic response can occur after a target site has been impacted by a single biologically effective unit of dose within a given time period (USEPA 2007b). The general form of the model is given by:

$$P(d) = 1 - e^{(-\lambda d)}$$

where P(d) is the probability of cancer from lifetime exposure at dose rated, and λ is a fitted dose coefficient (USEPA 2007a). Table 6.3 shows data for the target pharmaceuticals with sufficient tumor data to calculate SFs, and the resulting ADIs assuming an acceptable lifetime excess cancer risk of 1 in one million and that a person is exposed to the chemical at this dose for 365 days per year for 30 years over a 70 year lifetime.

Table 6.3
Tumor data used to calculate slope factors (SFs), acceptable daily intakes (ADIs), and drinking water equivalent levels (DWELs) for cancer endpoints for target pharmaceuticals

Drug	Study Design	Dose & Tumor Incidence (mg/kg-d)	Effect	Reference	Human Equiv. Doses (mg/kg-d)	SF (mg/kg-d)$^{-1}$	ADI* (mg/kg-d)	DWEL (µg/L)
Gemfibrozil	2 year, rat	0 = 1/50 30 = 8/50 300 = 17/50	Interstitial cell tumors of the testes in males	Fitzgerald et al. 1981	0 4.8 48	1.8E-03	0.0013	45
Phenytoin	2 year, mouse	0 = 5/48 50 = 14/49 160 = 30/50	Liver adenomas and carcinomas in females	NTP 1993	0 2.1 6.3	1.2E-03	0.00019	6.8

* Calculated assuming an acceptable lifetime excess cancer risk of 1 in one million and that a person is exposed to the chemical at this dose for 365 days per year for 30 years over a 70 year lifetime, or ADI = (10-6 × 25,550) / (SF × 10,950).

For drugs with evidence of carcinogenicity in animals but no available tumor incidence data to extrapolate SFs, a method proposed by Gaylor and Gold (Gaylor and Gold 1998) was used. Gaylor and Gold created the Carcinogenic Potency Database as part of the Cancer Potency Project at the Lawrence Berkeley Laboratory (Gold et al. 2005, Gold et al. 2006). The Carcinogenic Potency Database summarizes results from 6,153 chronic, long-term animal cancer tests on 1,485 chemicals, as published in the general literature through 1997 and by the National Cancer Institute/National Toxicology Program through 1998 (Gold et al. 2006). Gaylor and Gold (1998) reviewed the results of 2-year cancer bioassays for 139 chemicals tested by the National Toxicology Program (NTP) and determined that a "virtually safe dose" corresponding to a cancer risk of 1 in a million can be estimated by dividing a chemical's maximum tolerated dose (MTD) from 90-day studies in rodents by 740,000. The MTD is the highest dose predicted to produce minimal systemic toxicity over the course of a carcinogenicity study and is typically predicted from 90-day dose range finding studies (FDA 1995). Effects of concern include alterations in physiological function which could alter the animal's normal life span or interfere with interpretation of the carcinogenicity study, such as more than a 10% decrease in body weight gain relative to controls, target organ toxicity, or significant alterations in clinical pathological parameters. The MTD is usually the high dose selected for a carcinogenicity study (FDA 1995).

Table 6.4 shows ADIs for the target pharmaceuticals with evidence of carcinogenicity but no tumor data (atorvastatin, carbamazepine, risperidone, and simvastatin) calculated using this method. Table 6.4 also shows ADIs calculated using the MTD method for two other compounds (gemfibrozil and phenytoin) that have sufficient tumor data to calculate a slope factor by the standard method (see Table 6.3). These values are within a factor of about 3 of the values calculated using the standard method, demonstrating that this approach produces acceptable estimates of carcinogenicity potential for compounds without available tumor data.

Table 6.4
ADIs and DWELs calculated for target analytes with evidence of carcinogenicity but no tumor data, based on maximum-tolerated doses (MTDs)

Drug	MTD (mg/kg-d)	ADI* (mg/kg-d)	DWEL (µg/L)
Atenolol	1,500	0.0020	70
Atorvastatin	400	0.00054	19
Carbamazepine	250	0.00034	12
Gemfibrozil	300	0.00041	14
Phenytoin	36	0.000049	1.7
Risperidone	10	0.000014	0.49
Simvastatin	400	0.00054	19

* Calculated by dividing the MTD by a factor of 740,000.

DRINKING WATER EQUIVALENT LEVELS (DWELS)

For contaminants present in drinking water, ADIs can be converted to drinking water equivalent levels (DWELs) by making an assumption about the volume of water a person drinks in a day. As shown below, an ADI in mg/kg-d can be converted to a DWEL in nanograms per liter (ng/L; equivalent to a part per trillion or ppt) by multiplying it by an assumed body weight (*e.g.*, 70 kg, the U.S. EPA default adult body weight) and a metric unit conversion factor, and dividing by an average daily drinking water ingestion rate (*e.g.*, 2 liters per day)(USEPA 2006):

$$DWEL\,(ng/L) = \frac{ADI\,(mg/kg-d) \times 70\,kg \times 1{,}000{,}000\,ng/mg}{2\,L/d}$$

DWELs for carcinogens can be estimated assuming that a *de minimis* lifetime excess cancer risk of one additional cancer per one million lifetime exposures (10^{-6}) is acceptable, and that a 70 kg person consumes two liters of water daily (365 days per year) for 30 years over a 70 year lifetime (equivalent to 25,550 days) (USEPA 2006):

$$DWEL\,(ng/L) = \frac{10^{-6} \times 70\,kg \times 25{,}550\,d \times 1{,}000{,}000\,ng/mg}{SF\,(mg/kg-d)^{-1} \times 2\,L/d \times 30\,yr \times 365\,d/yr}$$

Tables 6.2, 6.3, and 6.4 show DWELs calculated for the target pharmaceuticals based on noncancer ADIs, SFs, and the MTD method, respectively.

Often, when setting action levels for contaminants in surface water systems, regulatory agencies incorporate an additional factor into the above equations to account for the possibility that on any given day, a person could be exposed to the substance through some other source than drinking water ingestion. This factor, called a relative source contribution or RSC, typically ranges in value from 20 to 80% based on knowledge about likely alternative sources of exposure to the chemical (USEPA 2000a; CADHS 2006; USEPA 2006). Thus, if this approach is used, the resulting DWEL would be 20 to 80% of the value based solely on ingestion of drinking water.

Comparisons to the Minimum Therapeutic Dose

The lower end of a drug's therapeutic range can represent a threshold for appreciable biological activity in target populations, and therefore could be considered a threshold for potential adverse effects (*i.e.*, a LOAEL). Table 6.5 shows minimum therapeutic doses and corresponding DWELs for each of the target analytes. When considering these values, an important consideration is that many drugs (and their corresponding therapeutic doses) are only approved for distinct population groups. For example, drugs in FDA Pregnancy Categories D and X may pose significant risks to the fetus and are not recommended for use during pregnancy. Other drugs have not undergone clinical testing in children such that minimum therapeutic doses can only be confidently applied to adults. Others are intended for use in life-threatening illnesses (*e.g.*, chemotherapy drugs, heart medications) such that a certain amount of risk of other health effects is expected and accepted, and effects other than the "therapeutic effect" may routinely occur at the minimum therapeutic dose.

For example, phenytoin has a minimum therapeutic dose in adults of 300 mg/day. This is equivalent to a daily dose of 4.3 mg/kg-day, assuming exposure to a 70 kg adult. If one divides this dose by a composite uncertainty factor of 1,000 (to account for extrapolation from a LOAEL to a NOAEL (10x), differences in sensitivity between adults and children (10x), and other assumed but unknown uncertainties in the available database (10x)), the resulting ADI would be 4.3 μg/kg-day. However, review of the toxicological database for phenytoin shows some effects at doses at or below the minimum therapeutic dose:

- Doses ranging from 20 to 50 mg/kg-day caused developmental effects in mice and rats (including reduced brain size and weight and brain cell degeneration) (Ohmori, Kobayashi, and Yasuda 1992, Ohmori et al. 1997, Hatta et al. 1999, Bittigau et al. 2002); see Appendix E). These would be equivalent to doses in humans ranging from 2.0 to 4.1 mg/kg-day (extrapolated using differences in body size between test animals and humans). If a standard UF of 3,000 (10 to account for extrapolation from animals to humans, 10 to account for extrapolation from a LOAEL to a NOAEL, 3 for intraspecies variation, 3 for extrapolation from subchronic to chronic, and 3 for database deficiencies) is applied to the LOAELs, the ADIs is would range from 6.7 to 17 μg/kg-day.
- A human equivalent dose of 2.1 mg/kg-day caused cancer in mice (NTP 1993). Using a "one-hit" model to calculate a cancer slope factor and assuming an acceptable lifetime excess cancer risk of 1 in 1,000,000, the resulting ADI is 0.19 μg/kg-day.

- These considerations show that some precautions should be taken when using the therapeutic dose in derivation of ADIs.

Table 6.5
Minimum therapeutic doses and equivalent drinking water concentrations for target pharmaceuticals

Drug	Minimum therapeutic dose (mg/d)	Age group	Assumed body weight (kg)	Endpoint	Minimum therapeutic dose (mg/kg-d)	Equivalent drinking water concentration (µg/L)
Atenolol	25	Adult	70	Hypertension	0.36	13,000
Atorvastatin	10	Child	30	Hypercholesterolemia	0.3	11,000
Carbamazepine	10	Child, <6	10	Epilepsy	1.0	35,000
Diazepam	2	Adult, geriatric	70	Anxiety	0.029	1,000
Diclofenac	100	Adult	70	Arthritis	1.4	49,000
Enalapril	2.5	Adult	70	Hypertension (w/ renal impairment)	0.036	1,300
Fluoxetine	10	Pediatric (children & adolescents)	30	Depression, obsessive compulsive disorder	0.33	12,000
Gemfibrozil	1,200	Adult	70	Lipid regulation	17	600,000
Meprobamate	200	Child, 6-12	30	Anxiety	7	250,000
Naproxen	125	Child	30	Juvenile arthritis	4	140,000
Phenytoin	300	Adult	70	Epilepsy	4.3	150,000
Risperidone	1.8	Adult	70	Schizophrenia	0.026	910
Simvastatin	10	Child, 10-17	50	Hypercholesterolemia	0.20	7,000
Sulfamethoxazole	400	Child (>2 mths)	30	Urinary tract infection	13	460,000
Triclosan	NA	NA	NA	NA	NA	NA
Trimethoprim	160	Adult	70	Urinary tract infection	2.3	81,000

CHAPTER 7
DEVELOPMENT OF RISK ANALYSIS METHODOLOGY FOR EDCS

INTRODUCTION

The toxicity assessment for EDCs focused on identifying data that could be used to update existing reference dose (RfDs) or acceptable daily intake (ADIs) for EDCs, or derive new ADIs, based on potential endocrine-mediated (estrogenic, androgenic, or thyroid-related) reproductive or developmental effects of EDCs in drinking water. Other potential endocrine mediated modes of action have been reported for EDCs (reviewed elsewhere, e.g., World Health Organization (WHO-IPCS 2002a)), but because the knowledge base for those modes of action is considerably smaller, they were not considered here. The approach presented is subject to change as the science of EDCs rapidly evolves.

In contrast to pharmaceuticals, widespread awareness and interest in the occurrence of EDCs in the environment began much earlier, as evidenced by the publication of the book *Silent Spring* in 1962. Many chemicals now known or believed to be EDCs were environmental contaminants of interest before their potential for endocrine-mediated effects was suspected. Thus, substantial toxicological databases (including RfDs or ADIs) already exist for many known or potential EDCs. Standard reproductive and developmental toxicity testing can identify some of the endocrine-mediated effects of EDCs; however, only within the last few years has research been directed extensively at developing and applying methods that identify not only the adverse effects but also the mechanisms by which they occur.

As discussed previously in Chapter 2, U.S. EPA's Endocrine Disruptor Screening Program (EDSP) Tier 2 bioassays are still under development. The resulting guidance (including standardized testing protocols) should enable risk assessors to evaluate whether specific chemicals are or are not EDCs and assess their importance to human health with greater ease and confidence. Furthermore, testing requirements can be expected to spur the development of a larger body of data that can serve as the basis for risk assessments. In the meantime, utilities are faced with questions from the public regarding the safety of drinking water containing trace concentrations of EDCs. Although some have quite reasonably decided to wait for further guidance from U.S. EPA regarding this complex situation, others have opted to take a proactive approach and wish to identify safe levels of EDCs in drinking water to serve as treatment targets. The following sections describe the methodology and data that were used to derive ADIs for the target EDCs based on endocrine-mediated endpoints.

METHODS

Approach

For most target EDCs, RfDs or ADIs were previously developed. A general description of the approach to identifying and selecting toxicity data to serve as the basis for ADIs and corresponding drinking water equivalent levels (DWELs) follows:

A search was conducted for previously published authoritative references describing oral RfDs, ADIs, tolerable daily intakes, permissible daily intakes, etc. First priority was given to the most recently published RfD used by the U.S. Environmental Protection Agency (U.S. EPA) to set drinking water criteria (federal maximum contaminant level (MCL)), presented in U.S.

EPA's Integrated Risk Information System (IRIS), or used in risk assessments for pesticide registrations, as described below. Other sources were values established by the California Office of Environmental Health Hazard Assessment (OEHHA), the Agency for Toxic Substances and Disease Registry (ATSDR) of the Centers for Disease Control, and international organizations.

Limited searches were conducted to identify No Observed Adverse Effect Levels (NOAELs) or Lowest Observed Adverse Effect Levels (LOAELs) for (preferably adverse) reproductive or developmental effects mediated through estrogenic, androgenic, or thyroid-related modes of action (MOAs). Emphasis was placed on data published since the establishment of the existing RfDs or ADIs. When they were available and applicable, secondary compilations and reviews were used as the major sources of information. These include reviews and reports published by WHO-IPCS (2002b); NTP-CERHR 2003, 2006, 2007; NRC (1999); and EC-BKH (2000). PubMed searches supplemented those sources of information (NRC 1999, NTP-CERHR 2003, NTP-CERHR 2006, NTP-CERHR 2007).

If the search identified at least one published NOAEL/LOAEL that met the criteria for selection and was less than the NOAEL/LOAEL used to derive the existing RfDs or ADIs, a new ADI that accounts for endocrine mediated effects was estimated using methods similar to those described for non-carcinogens[1] in Chapter 6. NOAELs/LOAELs of interest here describe doses associated with (preferably adverse) reproductive or developmental effects expected to be mediated through estrogenic, androgenic, or thyroid-related MOAs. Data describing other types of effects (e.g., recently published data on immunotoxicity, carcinogenicity, etc.) were not considered.

Sources and Selection of Previously Developed RfDs and ADIs

RfDs and ADIs are developed by a variety of national and international organizations and expert groups. For this project, RfDs and other types of ADIs documented in the following sources were considered:

U.S. EPA Integrated Risk Information System (IRIS). The U.S. EPA maintains IRIS, which includes reference doses and cancer slope factors developed by U.S. EPA for use in risk assessments. In some cases, drinking water criteria documents were used as the basis for IRIS records, particularly for chemicals that were regulated in 1990 or earlier (Donohue 2008). More recently, information in the IRIS records is used to develop the drinking water criteria.

U.S. EPA Drinking Water Criteria and Support Documents. U.S. EPA promulgates MCLs, which are federal regulatory limits on drinking water contaminants. As part of the process of setting an MCL for a drinking water contaminant, U.S. EPA estimates an RfD. Some of the early MCLs are supported by drinking water criteria documents that describe the development of the drinking water standard including the derivation of the RfD. In other cases, the drinking water criterion is based on the RfDs made available in IRIS. .

Pesticide Registration Decision Documents. To support registration processes, the U.S. EPA's Office of Prevention, Pesticides and Toxic Substances (OPPTS) conducts toxicological reviews and risk assessments for pesticides, to support registration processes. This process sometimes includes development of RfDs that differ from those in IRIS. These values are presented in Interim Reregistration Eligibility Decision (IRED), Reregistration Eligibility

1 Hormonally-mediated cancers (those related to endogenous estrogen exposure) currently are thought to occur only after the threshold for hormonal response is exceeded resulting in persistent overstimulation of estrogen-sensitive tissues (JECFA 2000).

Decision (RED), and Tolerance Reassessment Progress and Risk Management Decision (TRED) documents.

JECFA Evaluation of the Safety of Sex Steroids in Food. In 2000, the Joint FAO/WHO Expert Committee on Food Additives (JECFA) published a review and toxicological assessment of the safety of three endogenous hormones (17ß-estradiol, progesterone, and testosterone). The resulting toxicological monograph included a summary of the biology and toxicology of estradiol and its metabolites formed endogenously or ingested orally, emphasizing the safety of estradiol in foods.

NTP-CERHR Monographs. The National Toxicology Program's Center for the Evaluation of Risks to Human Reproduction (CERHR) recently completed reviews on the potential human reproductive and developmental effects of three of the target EDCs: butylbenzyl phthalate, DEHP, and bisphenol A.

California Office of Environmental Health Hazard Assessment (OEHHA) Public Health Goals (PHGs). The California Environmental Protection Agency (Cal/EPA) OEHHA has conducted its own risk assessments for certain drinking water contaminants, and presents the values as Public Health Goals (PHGs) in the form of water concentrations that are analogous to federal maximum contaminant level goal (MCLGs). In turn, they may derive state MCLs that are analogous to federal MCLs. In some cases, the California MCLs are more stringent than federal values. California PHGs/MCLs were identified for the reader's information as well as for comparison with the DWELs developed in the current report.

California OEHHA Proposition 65 Safe Harbor Levels. California OEHHA also implements the Safe Drinking Water and Toxic Enforcement Act of 1986 (Proposition 65). In that role, OEHHA develops no significant risk levels (NSRLs) for carcinogens and maximum allowable dose levels (MADLs) for chemicals that cause reproductive toxicity. The NSRL is the daily intake level calculated to result in one excess case of cancer in an exposed population of 100,000, assuming lifetime (70-year) exposure at the level in question. The MADL is the level at which the chemical is expected to have no observable adverse reproductive effect assuming exposure at 1,000 times that level. The NSRLs and MADLs are promulgated in part to assist interested parties in determining whether discharges to sources of drinking water are prohibited. (OEHHA-RCHAB 2008)

ATSDR MDLs. The Comprehensive Environmental Response, Compensation, and Liability Act (CERCLA) requires that the Agency for Toxic Substances and Disease Registry (ATSDR) develop jointly with U.S. EPA a list of hazardous substances most commonly found at facilities on the CERCLA National Priorities List (NPL); prepare toxicological profiles for each substance included on the priority list of hazardous substances; ascertain significant human exposure levels and the associated acute, subacute, and chronic health effects for hazardous substances; and assure the initiation of a research program to fill identified data needs related to the substances. As part of this mandate, ATSDR developed minimal risk levels (MRLs) for dozens of compounds, including some EDCs. The MRL is an estimate of the daily human exposure to a hazardous substance that is likely to be without appreciable risk of adverse noncancer health effects over a specified duration of exposure. MRLs were considered where available.

These documents and related databases summarizing information contained therein were reviewed to identify the toxic effects and effect level (e.g., NOAEL or LOAEL) used as the basis for the latest reference dose.

Selection of Data Describing Endocrine Mediated Toxicity

A limited literature search was conducted to compile data describing adverse endocrine-mediated effects published since the establishment of the latest suitable RfD or other ADI. Because the body of literature is extensive for some EDCs (e.g., bisphenol A), efforts were focused on review documents and identifying the lowest dose reported to cause an adverse endocrine-mediated effect in laboratory animals or humans. This value was then compared to the effect level used to develop the previously established RfDs or ADIs to assess whether inclusion of more recently published information on adverse endocrine-mediated effects might result in a substantially different recommended ADI. Because the science of endocrine disruption is rapidly evolving, it is likely that new data describing endocrine-mediated effects will continually become available.

Point of Departure and Uncertainty Factors

It is likely that experts will disagree about which effects of EDCs should be used as the point of departure (POD) for a risk assessment. For example, there is a lack of consensus regarding which effects are mediated through the endocrine system, which effects can be considered to be adverse, and whether non-adverse precursor effects should be used as a POD. However, U.S. EPA has already developed RfDs based on endocrine-mediated effects for two (and possibly more) chemicals, specifically perchlorate and atrazine (see below). Suitable data for development of ADIs for endocrine-mediated effects might result from human epidemiology, occupational health, or clinical studies, or studies of endocrine disruptive effects in laboratory animals that serve as surrogates for humans.

When studies of sufficient quality reporting adverse endocrine-mediated effects were identified in literature published since the latest existing RfD was established, the toxicological effects that occurred through an endocrine MOA in animals and humans were documented, as well as the dose levels at which these effects were observed. Data considered were limited to studies of in vivo effects in intact animals (e.g., studies involving ovariectomized animals—with ovaries surgically removed—were excluded). Corresponding dose-response data, including NOAELs and LOAELs, were tabulated. Uncertainty factors were then applied to these values to derive updated ADIs and DWELs.

Test Species. Data from studies with rats, mice, guinea pigs, rabbits, non-human primates, and humans were considered. Rodent species are most commonly used as laboratory animal surrogates for toxicological effects in human. Although mice can be more sensitive than rats, the rat is the most widely used species in routine toxicology testing (Jobling et al. 1998).

Route of Exposure. In general studies using an oral route of exposure are preferred for risk assessment of drinking water contaminants, and thus were used here. In some cases, chemicals administered orally are less active than those administered by other routes. For example, the endogenous estrogens are characterized by poor oral bioavailability and much greater activity when administered by other routes (e.g., intraperitoneal injection, implants, or transdermal application). Although the estrogen hormones are absorbed efficiently, they undergo rapid gut and hepatic metabolism (see Appendix H for more details). In other cases, EDCs require metabolic activation to exert their effects and thus are effective when ingested but not when they are administered by other routes.

Critical Effect. For a chemical to be classified as an EDC for the purposes of this project, its endocrine-mediated effects should occur at a dose smaller than that required to cause systemic toxicity. In other words, the endocrine-mediated effect should be the critical effect. For example, in mammalian toxicity testing, body weight loss of 10% or more is commonly used as an indicator of general systemic toxicity and frequently has been used a criterion for rejecting subjects from an ongoing study. However, caution must be used in the interpretation of body weight in studies of estrogenic chemicals because one of the classic effects of estradiol administration is reduced food intake and body weight (Goldman et al. 2000).

Endocrine-Mediated Effect. Although standard toxicity tests might detect effects that result from chemical interference with endocrine function, most standard toxicity tests are not designed to distinguish effects mediated through the endocrine system from effects occurring through other modes of action. In some cases, it might not be possible using available data to determine whether a reported effect is due to endocrine disruption. The budget and time-frame of the current project did not allow a detailed evaluation of the weight of evidence for endocrine-mediated effects *in vivo* and their corresponding modes of action for each selected chemical, so previous reliable assessments were used when these were available. In some cases, experts on the potential endocrine disruptive effects of the selected (potential) EDCs were contacted for guidance. However, in general, low-dose effects of known or potential EDCs on reproductive or developmental endpoints were considered to be likely to be mediated through the endocrine system for purposes of this assessment. This represents a conservative approach.

Adverse Effect. Both EDSTAC (1998) and NRC (1999) noted that experts on EDCs disagree about which effects of these chemicals can be considered to be adverse. Some uncertainty surrounds the distinction of adverse effects from biomarker or precursor effects (Ashby et al. 1997). For example, fetal death, cancer, or decreased fertility obviously are adverse effects, while the importance of small and isolated changes in the timing of first vaginal opening has been questioned (Ashby et al. 1997). Additional discussion of adverse effects related to specific bioassays is presented below.

Bioassays and Other Studies. In 1998, some of the Organization for Economic Cooperation and Development's whole life cycle or multi-generation tests with mammals evaluated many endpoints that are under endocrine control. Bioassays that are most suitable for evaluating EDCs are the one- and two-generation tests reproductive/ developmental toxicity tests, which include exposure during the prenatal period. Criteria evaluated during these tests include fertility, fecundity, sexual behavior, gametogenesis, ovulation, spermiation, implantation, development of specific tissues and organs, maintenance of pregnancy, parturition, growth and viability of offspring, lactation, and maternal behavior. All of these parameters are sensitive to endocrine-mediated toxicity. Additional endpoints that are recommended for evaluation are sperm production, sperm motility, accessory sex organ weight, daily vaginal cytology in females to evaluate the duration of the oestrus cycle, gonadal histopathology, and sex steroid production. The two-generation tests are the most rigorous, allowing evaluation of growth, development, and sexual function of the F1 generation and monitoring of the F2 generation through to weaning. Endpoints recommended for examination in offspring (F1 and F2) that have been exposed in utero include timing of vaginal opening, testicular descent, anogenital distance, nipple retention, accessory sex organ weight, and vaginal cytology. (Jobling 1998)

Uterotrophic assays are short term in-vivo screening bioassays that use changes in uterine weight gain and development in immature animals or in ovariectomized adults as indicators of estrogenic effects. The uterotrophic assays conducted with intact immature animals were developed as short-term screening tests but may be used for risk assessment. Those conducted in ovariectomized (non-intact) animals are used only as screens to investigate activity and mode of action. The removal of the gonads and the sex steroids they produce alters the normal internal hormonal milieu and does not allow for assessment of effects that occur through the hypothalamic-pituitary-gonadal (HPG) axis (EDSTAC 1998), so the results of bioassays with ovariectomized or castrated animals cannot be considered to be representative of effects in intact animals. According to U.S. EPA's Endocrine Disruptor Screening and Testing and Advisory Committee (EDSTAC), one endpoint, uterine gland development, examined in the developmental (immature rodent) uterotrophic assay is considered to be an adverse effect. Uterine gland development is a classical developmental toxicity endpoint; there is a defined ontogenic pattern, a period of sensitivity, and an irreversible, adverse outcome (glands are absent or their numbers greatly reduced in adults following treatment between 10 and 14 days after birth, or at postnatal day (PND) 10-14) (EDSTAC 1998). Uterine weight gain and epithelial hypertrophy (measured as cell height) are reversible (EDSTAC 1998), and thus might not represent an adverse effect.

Although the pubertal male assay was developed only as a screening assay, the U.S. EPA has previously used measures of altered puberty (e.g., a 2-day delay in preputial separation caused by vinclozolin or a similar delay in vaginal opening in the female) as adverse effects in risk assessments.

Work conducted on the effects of estrogenic chemicals on the prostate gland in mice has gained a great deal of attention. According to vom Saal et al. (vom Saal and Hughes 2005):

- ...prostate gland epithelial hyperplasia, associated with an increase in prostate size and a permanent increase in prostate androgen receptors, is induced in CF-1 male mouse fetuses by a very small increase relative to controls of about 50 pg/ml total serum estradiol (estradiol was administered to pregnant females via Silastic capsules (Dow) implanted subcutaneously; this corresponds to an increase of approximately 0.1 pg/ml free (bioactive) serum estradiol... because only a very small proportion (0.2%) of the total circulating estradiol in mouse fetuses is free (unconjugated and not bound to plasma proteins). A permanent increase in prostate androgen receptors will increase the response of the prostate to stimulation by androgen throughout life, and this is a risk factor for prostate disease during aging in men (Richter, Timms, and Saal 2004).

In some cases, non-adverse precursor effects may be used as the POD for U.S. EPA risk assessments. For example, suppression of leutenizing hormone (LH) in female Sprague-Dawley rats was used as the POD for a recent risk assessment conducted for a reregistration eligibility decision for atrazine (USEPA-OPPTS 2006a). In this rat strain, LH suppression ultimately results in the development of mammary tumors, an effect that is not expected to occur in human females. However, U.S. EPA determined that LH suppression might lead to other types of effects in the human female and selected that endpoint as the POD. The selection of LH suppression as the POD has important implications for the regulation of other EDCs. LH suppression in itself is not necessarily an adverse effect, but if the magnitude and duration of change are sufficient, LH suppression could lead to adverse effects. U.S. EPA has chosen to base its risk assessment on the precursor effect. Similarly, U.S. EPA selected a non-adverse precursor effect as the POD in developing its RfD for perchlorate (USEPA-IRIS 2005).

The case of perchlorate illustrates the contentious nature of risk assessments for EDCs. Perchlorate is an EDC that acts on the thyroid by competitively inhibiting the uptake of iodide into the thyroid gland. Iodide is required for thyroid hormone synthesis, so a chemical that inhibits iodide uptake to a sufficient extent can interfere with thyroid hormone production. Thyroid hormone levels are homeostatically controlled; i.e., the body possesses compensatory mechanism to keep thyroid hormone levels within a set range. When thyroid hormone levels decline, the body increases production of thyroid stimulating hormone (TSH) to boost thyroid hormone production. If these mechanisms are overwhelmed, hypothyroidism (deficient thyroid hormone production) occurs, leading to a plethora of adverse health effects. Adverse effects typically serve as the basis for the POD in the development of an RfD. It is generally agreed that short-term, reversible changes in endpoints such as enzyme levels are not considered to be adverse effects for the purposes of risk assessment. In the case of perchlorate, inhibition of iodide uptake is not considered to be an adverse effect but rather the first step (precursor event) in a progression of events that can ultimately result in adverse effects, but only if compensatory mechanisms are overcome.

Because of controversy surrounding the development of an RfD for perchlorate, the U.S. Department of Defense (DOD), U.S. Department of Energy (DOE), the National Aeronautics and Space Administration (NASA), and the U.S. EPA charged the National Academy of Sciences National Research Council (NRC) with assessing the potential adverse health effects of perchlorate ingestion (NRC 2005). The NRC convened an expert committee to assess the current state of science regarding potential adverse effects of perchlorate ingestion. The committee was also asked to determine whether U.S. EPA's findings in its 2002 draft risk assessment were consistent with current scientific evidence. Initially, U.S. EPA defined changes in serum levels of thyroid hormone and TSH as adverse effects. The NRC committee disagreed, finding that transient changes in serum levels of these hormones are not adverse effects, that hypothyroidism (deficient thyroid hormone production) is the first adverse effect that can occur after perchlorate ingestion, and that a prolonged decrease in iodide intake is necessary to cause an adverse effect (NRC 2005).

Initially, U.S. EPA proposed changes in brain structure, thyroid micropathology, and serum thyroid hormone levels in rats as the point of departure for the reference dose for perchlorate (NRC 2005). Based on the recommendations of the NRC committee, U.S. EPA instead based its point of departure on inhibition of iodide uptake in humans. As discussed above, the point of departure usually is an adverse effect, but in this case, U.S. EPA used a more conservative precursor effect. Inhibition of iodide uptake must precede any adverse effect, but the adverse effects are not an inevitable result of inhibition of iodide uptake (NRC 2005).

Uncertainty Factors. Standard uncertainty factors are described in Chapter 9. In addition, U.S. EPA also applies an additional 10× safety factor mandated by the 1996 Food Quality Protection Act for chemicals of special concern for children (Gray and Foster 2003). For example, this safety factor was retained in the recent U.S. EPA risk assessment for vinclozolin (USEPA 2000).

RESULTS

Toxicity data identified in the literature search and considered in the derivation of updated ADIs are described in Appendix H for each target EDC. Table 7.1 summarizes the studies selected as the point of departure for each EDC, and the corresponding ADIs and DWELs.

Table 7.1

Toxicological data used to develop acceptable daily intakes (ADIs) and drinking water equivalent levels (DWELs) for EDCs

Compound	Study duration	Effect dose (mg/kg-d)	Effect	Reference	UF (animal to human)	UF (intra-species)	UF (LOAEL to NOAEL)	UF (subchronic to chronic)	UF (dbase)	Composite UF	ADI (μg/kg-d)	DWEL (ppb)
Atrazine	Multi-Generation	0.5 (LOAEL)	Developmental, rat	U.S. EPA MCL	10	10	1	1	1	5,000[1] (includes UF of 50 for cancer)	0.1	3 (same as US EPA MCL)
Bisphenol A	3 generation	5 (NOAEL)	Developmental, rat	Tyl et al., 2002	10	10	1	1	1	100	50	1,800
Butylbenzyl phthalate	2 generation	100 (LOAEL)	Reproductive/ developmental, rat	Aso et al. 2005	10	3	10	3	1	1,000	100	3,500
Diethylhexyl phthalate	GD6-PND21	1.215 (NOAEL)	Reproductive/ developmental, rat	Andrade, Grande et al. 2006a	10	3	1	3	1	100	12	420
17ß-Estradiol	Therapeutic dosing	0.005 (NOEL)	Endocrine, human	JEFCA, 2000	1	10	1	3	3	100	0.050	1.8
Estrone	2 wk	0.004 (NOAEL)	Endocrine/ liver, human	Mashchak et al. 1982	1	10	1	10	3	300	0. 013	0.46
Ethynylestradiol	Therapeutic dosing	0.0001 (LOAEL)	Endocrine, human	Mashchak et al. 1982	1	10	10	3	3	1,000	0.00010	0.0035
Lindane	2 generation	0.56 (LOAEL)	Systemic (reproductive at higher doses), rat	Matsuura et al. 2005	10	3	10	3	1	1,000	0.56	20
Linuron	2 year	0.625 (LOEL)	Hematological, dog	du Pont 1962 as cited in U.S. EPA-IRIS 1993	100	1	3	1	1	300[2]	2.0 (same as U.S. EPA RfD)	70
Methoxychlor	GD11-17	0.020 (LOAEL)	Developmental, mouse	Palanza, Morellini et al. 2002	10	3	10	3	1	1,000	0.020	0.70

(continued)

Table 7.1 (Continued)

Compound	Study duration	Effect dose (mg/kg-d)	Effect	Reference	UF (animal to human)	UF (intra-species)	UF (LOAEL to NOAEL)	UF (subchronic to chronic)	UF (dbase)	Composite UF	ADI (μg/kg-d)	DWEL (ppb)
4-Nonylphenol	3 generation	1.5 (NOAEL)	Developmental, rat	Tyl et al., 2006	10	3	1	1	1	30	50	1,800
Octylphenol	3 generation; PND -5	15 (LOAEL)	Developmental, rat	Tyl et al., 1999; Nagao et al., 2001	10	3	1	3	1	100	150	5,300
Vinclozolin	"chronic"	1.2 (NOAEL)	Systemic, dog	U.S. EPA 2000 RED	10	3	1	1	3	100	12 (same as U.S. EPA FQPA RfD)	420

[1] According to current U.S. EPA (2002) guidance, a UF of 3 should be applied for the intraspecies uncertainty factor since the critical effect is based on a known sensitive population group (e.g., pups), and a UF of 3 should be applied for the study duration uncertainty factor the critical study is a reproductive/ developmental study (see Table 6.1); however, a factor of 10 for the intraspecies UF and no factor for the study duration UF was retained to be consistent with the MCL.

[2] According to current U.S. EPA (2002) guidance, a UF of 3 should be applied for the intraspecies uncertainty factor since the critical effect is based on a known sensitive population group (e.g., pups), and a UF of 3 should be applied for the study duration uncertainty factor the critical study is a reproductive/ developmental study (see Table 6.1); however, a factor of 10 for the intraspecies UF and no factor for the study duration UF was retained to be consistent with the MCL.

CHAPTER 8
RELATIVE EXPOSURE

Public concerns are often raised when "intersexed" male fish in local lakes and rivers are reported to be "growing eggs" or demonstrating other reproductive anomalies. These concerns revolve around whether the "abnormal" fish are a result of human-made chemical agents that have been released into the environment. Recent analyses have detected estrogenic and androgenic chemicals in surface water systems, as well as other types of compounds known to be biologically active such as pharmaceutical agents. These detections raise even more questions: Are these chemicals responsible for causing the abnormalities in fish? Can they also be found in our drinking water? How much might we be exposed to, and what are the risks to human health? A comparative or relative risk assessment offers a useful tool to help interpret data on pharmaceuticals or endocrine disrupting compounds detected in drinking water sources, and accurately address the questions from those in the community.

Comparative risk assessment is a tool that can inform risk ranking, prioritization, and relative exposure characterization. In one of the first assessments of this type, Belzer et al. (2004) determined typical dietary exposures to nitrate (a chemical that is naturally found in the environment and also a component of fertilizers and other agricultural products) and compared them with drinking water exposures to perchlorate (a naturally found chemical that is also manufactured for use as an oxidizer for solid rocket engines, automobile airbags, fireworks, flares, and some fertilizers). Certain waste handling practices in the past by some industries or government agencies have led to the release of perchlorate into ground and surface waters.

What makes these two chemicals candidates for a comparative risk assessment is that they both can affect the thyroid gland via the same mechanism of action (known as iodide uptake inhibition (Greer et al. 2002). Although the U.S. EPA has set an "acceptable exposure" level or reference dose (RfD) for nitrate, it is based on a different, and rather overt, toxicological effect (methemoglobinemia or "blue-baby syndrome") that occurs at a very high level of exposure, whereas the current RfD for perchlorate is based on a no observed effect level (NOEL) at which no inhibition of iodide uptake, the biochemical precursor to any thyroid effects, occurs (NRC 2005).

Belzer et al. (2004) conducted an assessment of the two agents, comparing the relative potential of each agent to cause iodide uptake inhibition at typical levels of exposure associated with background levels in the environment. Although perchlorate is more potent than nitrate for the inhibition of iodide uptake, many common foods contain substantial quantities of nitrate compared to the environmental levels of perchlorate. The results of this assessment demonstrated that a healthy diet has quantities of natural nitrate that are predicted to greatly overshadow any effect of perchlorate from food or drinking water on inhibition of iodide uptake into the thyroid, yet no overt thyroid effects from these background levels of nitrate are seen in the general populace. Thus, using a comparative risk assessment, Belzer et al. (2004) was able to show that environmental levels of perchlorate are not likely to present a significant health risk.

The same process can be applied to chemicals suspected of being endocrine disruptors that act via the same mode of action or physiologic mechanism. In the project presented herein, we detected several chemicals with recognized estrogenic, androgenic, or thyroid-active modes of action in finished drinking water. Some of these agents were also detected in food; thus, like perchlorate and nitrate, one can conduct an examination of relative exposure or conduct a comparative risk assessment.

In this chapter, information from this project and the general literature is used to develop a simple example of a relative or comparative risk assessment. Conducting a fully developed assessment is beyond the scope of this project; however, the examples shown here provide enough useful information to illustrate the utility of this assessment tool. To simplify the findings, the *maximum* detected concentration in the waters is used as a worst-case scenario. Thus, the results presented herein are clearly conservative given that all other values of these compounds at the various facilities are lower. The maximum detected levels in foods and water sources are noted using estradiol equivalents (EEq); this assay does not detect individual chemicals per se, but measures an integrated cellular response to estrogenic chemicals. For comparison, the literature was examined for endogenous concentrations of estrogenic compounds in cow and human breast milk.

As noted in earlier chapters, several potential EDCs were detected in waters from the utilities that were part of this study. In Table 8.1, the maximum detected concentrations of potential EDCs in source, raw intake, finished drinking, and distribution waters are summarized. For instance, nonylphenol was detected in source, raw, and finished drinking water at values of 630, 130, and 100 ng/L, respectfully. Bisphenol A was detected with a maximum detection in each of the three types of waters tested, at values 120, 14, and 25, ng/L, respectfully. The most potent of the estrogenic compounds of the agents that were detected, estradiol (see Figure 5.1) was detected at a lower maximum concentration in source water (2 ng/L) and raw intake water (17 ng/L). The compounds detected at higher concentrations (e.g., nonylphenol) are approximately one thousand times less potent than estradiol. For a brief summary of potency, see the footnote below.[2]

Not only were estrogens and suspected estrogenic compounds detected in some utility water sources, but they have also been detected in foods using the cellular bioassay. Using the highest maximum value for source, raw, finished, and distribution water, the results show that source water had the highest response at 4.8 EEq ng/L and raw water was about half that value at 2.1 EEq ng/L. Finished and distribution water reported no activity above the MRL. The highest maximum value for source water shows approximately the same level of estrogenic activity as a serving of green tea, which has nearly one-fourth the estrogenicity of a serving of coffee. The highest maximum activity per serving was for soy milk and soy baby formula. Values for these are approximately 840 and 396 fold higher than the maximum for source waters. This demonstrates that some foods on a per serving basis have considerably greater levels of estrogenic activity.

Not only can we compare various types of water to common foods for estrogenic activity, we can also compare selected estrogens in water (or food) to endogenous estrogen levels. The maximum concentrations in source and raw water intake are 6.2 and 0.94 ng/L for estrone and 2 and 17 ng/L for estradiol. These chemicals were not detected in finished or distribution water. For human breast milk, estradiol mean values are approximately 36 to 45 ng/L (reported as 0.036 to 0.045 µg/L) (Messripour, Forooghi-Abary, Dashti-Ardakani 2002). Mean values for 17ß-estradiol are 3.4 to 5.5 ng/L (reported as 0.8 to 1.3 ng/cup) and mean values for estrone are 11.4 to 33.8 ng/L (reported as 2.7 to 8.0 ng/cup) (Wolford and Argoudelis 1979). Cow's milk appears to have similar general concentration ranges of estrogens as human breast milk. The mean values for 17ß-estradiol range from 3.5 to 60 ng/L (Wolford and Argoudelis 1979; Harmann et al.

[2] As presented In Figure 5.1 and 5.2, the data show in E-screen tests potency as follows: estradiol >>> coumestrol > equol ~ genistein > biochaninA ~ diadzein >naringenin > formononetin > apigen; and in comparing estrogen to phytoestrogens in E-screen tests, we see: estradiol ~ ethynyl estrodiol >estone>>octylphenol ~ nonylphenol ~ bisphenol A > butylbenzylphtalate.

1998). The mean values for estrone range from 9.1 to 260 ng/L (Wolford and Argoudelis 1979; Fritsche and Steinhart 1999; Harmann et al. 1998).

Clearly, unanswered questions persist about the potential effects of unintentional exposure to estradiol, estrone, and ethynylestradiol. However, although these chemicals are commonly detected in wastewater treatment plant effluent, wastewater influenced surface water, and occasionally in source or raw water, they have rarely been detected in finished drinking water and have not been reported to occur in drinking water in the U.S. in either the current or the previous literature. Previous studies have shown that these phenolic steroids can be readily removed during drinking water treatment processes, such as chlorination and ozonation (Snyder et al. 2007a). Kuch and Ballschmiter monitored steroids in drinking water samples from three treatment plants (10 samples total) in Germany (Kuch and Ballschmiter 2001). They detected estrone in four of the ten samples with concentrations from 0.2-0.6 ng/L. Aherne and Briggs reported detecting ethynylestradiol in finished drinking water from the United Kingdom in the range of <1 to 4 ng/L using an immunoassay technique (Aherne and Briggs 1989). Kolpin et al. (2002) reported ethynylestradiol and estradiol levels in U.S. streams and rivers as large as 831 and 200 ng/L, respectively. However, some of these data were later retracted due to analytical error (http://toxics.usgs.gov/regional/est_errata.html).

Even if trace concentrations of estradiol or estrone were ingested in drinking water, it appears unlikely to be a significant route of exposure compared with other sources. Endogenous production is the greatest route of exposure for most people. Prepubertal children are noted as a potentially sensitive subpopulation because their endogenous estrogen levels are naturally very low. However, people in the U.S., including infants and prepubertal children, typically are exposed to quantities of estradiol, estrone, and other hormones normally found in human breast milk, cow's milk, and other animal products (see Tables 8.2 and 8.3) that exceed doses or exposures normally received through ingestion of drinking water. Clearly more research is needed to adequately compare and contrast estrogens and suspected estrogens in water and food.

Table 8.1
The maximum detected concentrations of potential EDCs in utility water (ng/L)*★

Compound	Source Water	Raw Intake Water	Finished Drinking Water	Distribution Water
Nonylphenol	630	130	100	110
Bisphenol A	120	14	25	★
Octylphenol	68	★	★	★
Butylbenzyl phthalate	60	54	★	★
Formononetin	22	6.1	★	★
Naringenin	17	2.6	★	24
Genistein	9.5	4.3	2.9	★
Daidzein	9.2	5.6	5.8	★
Biochanin A	8.3	1.5	★	★
Estrone	6.2	0.94	★	★
Progesterone	3.1	3.1	0. 57	★
Estradiol	2	17	★	★
Chrysin	1.4	2.1	★	★
Coumestrol	2.0	2.0	★	★
Ethynylestradiol	★	1.4	★	★
Testosterone	★	1.2	★	★

* This data was obtained from the following tables: Table 3.2: Results for source water; Table 3.3: Results for raw intake water; Table 3.4: Results for finished drinking water; Table 3.5: Results for distribution water. If data results were non-detect, the data was ignored.

★ The data for these cells were assumed to be non-detect, per chapters 3 and 5.

Table 8.2
Maximum estradiol equivalent (EEq) results in utility water and food and beverage products*

Description of Food and Beverage Products	**Maximum EEq (ng/L)***
Soy Milk	4200
Soy Baby Formula	1900
Soy Sauce	510
Beer	140
Coffee	17
Green Tea	4.4
Vegetable Juice	3.3
Baby Formula	0.85
Milk	0.81
Apple Juice	0.79
Bottle Water	<0.16
Description of Water Utilities	
Source (n=18)	4.8
Raw (n=17)	2.1
Finished (n=16)	0.77
Distribution (n-15)	0.20

* This data was obtained from the following tables: Table 5.1: Estradiol equivalent results in utility water; Table 5.3: Estradiol equivalent results in food and beverage products. If data results were non-detect, the data was ignored.

Table 8.3
Concentrations of estrogenic hormones in dairy cow milk and human breast milk

Milk Type	Estrogen	Concentration	Reference
Human	Unspecified	15 – 840 μg/L	Borgert et al. 2003, Hamosh 2001
Human	Estradiol	0.036 μg/L (mean) 0.045 μg/L (mean)	Messripour, Forooghi-Abary, and Dashti-Ardakani 2002
Human	17ß-Estradiol	0.8 – 1.3 ng/cup	Wolford and Argoudelis 1979
Human	Estrone	2.7 – 8.0 ng/cup	Wolford and Argoudelis 1979
Cow	Unspecified	≤ 0.017 μg/L	Lopez et al. 2004
Cow	17ß-Estradiol	0.01 μg/L	Harmann et al. 1998
Cow	17ß-Estradiol	0.01 – 0.09 μg/L*	Fritsche and Steinhart 1998, Harmann et al. 1998
Cow	17ß-Estradiol	0.01 - 0.06 μg/L	Harmann et al. 1998
Cow	17ß-Estradiol	10.3 ng/L (mean) 14.4 ng/L 5.0 ng/L (mean) 9.0 ng/L (mean) 6.4 ng/L (mean) 3.5 ng/L (mean)	Wolford and Argoudelis 1979
Cow	Estrone	0.01 – 0.26 μg/L†	Fritsche and Steinhart 1998, Harmann et al. 1998
Cow	Estrone	0.03 – 0.12 μg/L	Harmann et al. 1998
Cow	Estrone	9.1 ng/L (mean) 20.2 ng/L (mean) 33.7 ng/L (mean) 10.0 ng/L (mean) 55.8 ng/L	Wolford and Argoudelis 1979

* Range of mean concentrations (detected levels only) provided in a review of hormones in cow's milk, including skim, reduced fat, and whole milk and both processed and unprocessed milk.

† Range of mean concentrations provided in a review of hormones in cow's milk, including skim, reduced fat, and whole milk and both processed and unprocessed milk.

CHAPTER 9
COMPREHENSIVE RISK ANALYSIS

Table 9.1 compares the maximum-detected concentrations in either finished drinking water or distribution drinking water for each of the target pharmaceuticals (see Chapter 3) to calculated drinking water equivalent levels (DWELs). The maximum-detected concentrations shown here are the single highest discrete sample concentrations, to provide a conservative "worst-case" scenario approach. The maximum values reported in Table 9.1 may differ slightly from the highest detected values provided in Chapter 3, which reports the maximum of averaged duplicate and triplicate samples.

Nine of the 20 compounds and their metabolites were detected in finished drinking water. An additional two compounds (norfluoxetine and risperidone) were also detected in distribution drinking water. The remaining nine compounds were not detected in either finished or distribution drinking water at detection limits of 2.5 ng/L or lower. Of the 11 pharmaceuticals that were detected, three compounds (the two antibiotics/ antibacterials, sulfamethoxazole and triclosan, and the anti-anxiety agent meprobomate) have calculated DWELs greater than 200 µg/L and therefore appear to be of relatively low toxicity, although available toxicity data of sufficient quality to develop the acceptable daily intake (ADIs) were limited to three or fewer studies for sulfamethoxazole and meprobomate (see Appendix E). Minimum margins of safety (i.e., the DWEL divided by the maximum-detected finished or distribution water concentration) for these compounds were 6,000 for meprobamate, 6,000,000 for sulfamethoxazole, and 2,200,000 for triclosan. The estimated potential hazard associated with exposure to the beta-blocker atenolol is also relatively low, with a DWEL of 70 µg/L and a minimum margin of safety compared to the maximum-detected finished or distribution water concentration of 2,700. The margins of safety of three other detected compounds, diazepam (110,000), fluoxetine (41,000) and norfluoxetine (44,000), are also very large.

The four remaining compounds that were detected in finished or distribution water (carbamazepine, gemfibrozil, phenytoin, and risperidone) show some evidence of carcinogenicity in animal studies, and calculated ADIs and DWELs were all based on cancer endpoints. Of these, the atypical anti-psychotic risperidone appears to be potentially the most potent carcinogen, with reported increases in pituitary gland adenomas in female mice, endocrine pancreatic adenomas in male rats, mammary gland adenocarcinomas in female mice and rats and male rats, and renal tubular adenomas and adenocarcinomas in male rats (see Appendix E). Risperidone was reported as detected at a level above the reporting limit (2.5 ng/L) in only one of 15 distribution water samples, and no finished drinking water samples. The minimum margin of safety was 170. The remaining three compounds with evidence of carcinogenicity (carbamazepine, gemfibrozil, phenytoin) were detected in one-quarter or more of the finished and distribution drinking water samples; corresponding minimum margins of safety were 670 for carbamazepine, 21,000 for gemfibrozil, and 210 for phenytoin.

The assumed cancer potency of carbamazepine applied in this assessment is uncertain because, while carbamazepine is identified as showing evidence of carcinogenicity in rats by Singh et al. (Singh, Driever, and Sander 2005) and on its U.S. Food and Drug Administration (FDA) label (Novartis 2000), data on tumor incidence per dose were not identified for use in this assessment and the carcinogenicity potential was estimated using the MTD approach. The estimated cancer potency of gemfibrozil is also uncertain because the assumed mechanistic course leading to cancer (proliferation of peroxisomes in the liver) occurs in rats but has not been observed in human liver, and numerous epidemiologic and clinical studies of gemfibrozil found no excess cancer mortality in humans (IARC 1996a, IARC 1996b).

Table 9.1
Comparison of pharmaceutical DWELs with maximum drinking water concentrations

Drug	Class	ADI (μg/kg-d)	DWEL (μg/L)	Maximum finished water conc. (μg/L)	Minimum margin of safety
Atenolol	Beta-blocker	2.0†	70	0.026	2,700
Carbamazepine	Anticonvulsant	0.34†	12	0.018	670
Diazepam	Benzodiazepine tranquilizer	1.0*	35	0.00033	110,000
Fluoxetine	SSRI antidepressant	0.97‡	34	0.00082	41,000
Norfluoxetine	Metabolite	0.97‡	34	0.00077**	44,000
Gemfibrozil	Antilipidemic	1.3‡	45	0.0021	21,000
Meprobamate	Antianxiety agent	7.5*	260	0.043	6,000
Phenytoin	Anticonvulsant	0.19‡	6.8	0.032	210
Risperidone	Antipsychotic	0.014†	0.49	0.0029**	170
Sulfamethoxazole	Anti-infective	510*	18,000	0.0030	6,000,000
Triclosan	Antibacterial	75*	2,600	0.0012	2,200,000
Atorvastatin	Antilipidemic	0.54†	19	<0.00025	>76,000
o-hydroxy atorvastatin	Metabolite	0.54†	19	<0.00050	>38,000
p-hydroxy atorvastatin	Metabolite	0.54†	19	<0.00050	>38,000
Diclofenac	NSAID	67*	2,300	<0.00025	>9,200,000
Enalapril	ACE inhibitor	0.23*	8.1	<0.00025	>32,000

(continued)

Table 9.1 (Continued)

Drug	Class	ADI (µg/kg-d)	DWEL (µg/L)	Maximum finished water conc. (µg/L)	Minimum margin of safety
Naproxen	NSAID	570*	20,000	<0.00050	>40,000,000
Simvastatin	Antilipidemic	0.54†	19	<0.00025	>76,000
Simvastatin hydroxy acid	Metabolite	0.54†	19	<0.00025	>76,000
Trimethoprim	Antibacterial	190*	6,700	<0.00025	>27,000,000

* Single highest discrete sample concentration, from finished drinking water unless otherwise noted.
† Derived from a cancer endpoint using the Maximum Tolerated Dose (MTD) method.
‡ Derived from a noncancer endpoint.
§ Derived from a cancer endpoint using the tumor data and the one-hit model.
**Concentration from distribution drinking water.

Several agencies consider phenytoin a possible or likely human carcinogen. The International Agency for Research on Cancer (IARC 1996c) lists phenytoin as a Group 2B carcinogen (possibly carcinogenic to humans) based on sufficient evidence in animals and inadequate evidence in humans. The National Toxicology Program (NTP) lists phenytoin as a "Suspect Human Carcinogen" based on evidence of the carcinogenicity of phenytoin in mice and rats but equivocal evidence in humans (NTP 2003). Based on the IARC assessment, the State of California also lists phenytoin as "Known to Cause Cancer" under Proposition 65, the Safe Drinking Water and Toxic Enforcement Act of 1986 (CalEPA 2006).

RISK ANALYSIS OF EDCS

Table 9.2 compares the maximum detected finished or distribution water concentrations of target EDCs (see Chapter 3) to calculated DWELs. Only four of the 13 compounds and their metabolites were detected in finished or distribution drinking water in the current study. The remaining nine compounds were not detected above their reporting limits.

Table 9.2
Comparison of EDC DWELs with maximum drinking water concentrations

Chemical	Class	ADI (µg/kg-d)	DWEL (µg/L)	Maximum finished water conc. (µg/L)	Minimum margin of safety
Atrazine	Herbicide	0.1	3	1.0	3
Bisphenol A	Industrial chemical	50	1,800	0.025	72,000
Linuron	Herbicide	2.0†	70	0.0083	8,400

(continued)

Table 9.2 (Continued)

Chemical	Class	ADI (μg/kg-d)	DWEL (μg/L)	Maximum finished water conc. (μg/L)	Minimum margin of safety
4-Nonylphenol	Surfactant, synthesis intermediate, nonylphenol ethoxylate degradate	50	1,800	0.11‡	16,000
Butylbenzyl phthalate	Phthalate plasticizer	100	3,500	<0.050	>70,000
Diethylhexyl phthalate	Phthalate	12	420	<0.10	>4,200
17ß-Estradiol	Endogenous estrogenic steroid hormone, drug	0.050	1.8	<0.00050	>3,600
Estrone	Endogenous estrogenic steroid hormone, drug	0.013	0.46	<0.00020	>2,300
Ethynylestradiol	Pharmaceutical estrogen	0.00010	0.0035	<0.0010	>3.5
Lindane	Organochlorine pesticide	0.56	20	<0.010	>2,000
Methoxychlor	Organochlorine pesticide	0.020	0.70	<0.010	>70
Octylphenol	Surfactant, chemical synthesis intermediate, degradate of octylphenol ethoxylates	150	5,300	<0.025	>210,000
Vinclozolin	Agricultural fungicide	12†	420	<0.010	>42,000

* Single highest discrete sample concentration, from finished drinking water unless otherwise noted.
† U.S. EPA RfD
‡ Concentration from distribution drinking water.

Of the four EDCs that were detected, none were present at concentrations above their DWELs. Atrazine has the lowest DWEL, with a value of 3 μg/L. The maximum detected concentration in finished drinking water was 1 μg/L, yielding a minimum margin of safety of 3. Two compounds (bisphenol A and 4-nonylphenol) have calculated endocrine-based DWELs greater than 200 μg/L and therefore appear to be of relatively low concern for this toxicity endpoint. Minimum margins of safety for these compounds were 72,000 for bisphenol A and 16,000 for 4-nonylphenol. The toxicity evaluation also indicates that the herbicide linuron is of relatively low concern for endocrine-mediated endpoints; the DWEL for linuron was 70 μg/L with a minimum margin of safety of 8,400.

UNCERTAINTIES

Uncertainty refers to a lack of knowledge, which is not the same as variability. For example, in the field of toxicology, studies evaluating the potential for adverse effects to the exact populations and scenarios of interest (e.g., humans, particularly children, exposed to low levels of multiple compounds through multiple exposure routes throughout their daily activities) are rarely available, and therefore significant uncertainties can exist about how available data correspond to actual human exposures. When deriving ADIs in this assessment, an effort was made to account for uncertainties in the data available and how they might be applied to potentially exposed populations, with the goal being to err on the side of conservatism.

The very young (including the fetus) and the elderly are generally most susceptible to adverse effects from exposure to chemical compounds. The young are more susceptible because they are rapidly developing and younger cells are more easily harmed than mature cells, their bodies may differ in how they absorb metabolize, and excrete compounds, and their smaller size means that a given exposure could be greater on a per body weight basis (USEPA 2000b). The elderly have increased sensitivity because of reduced ability to metabolize agents and compensate for effects (USEPA 2001). Gender, genetic background, and health status can also affect how a person responds to an exposure. In this assessment dose-response data from the studies that showed adverse effects at the lowest doses, often corresponding to reproductive effects in females or developmental effects in the fetus or offspring, were considered in developing ADIs, and uncertainty factors were applied to account for potentially sensitive subpopulations. Therefore, estimated ADIs and DWELs incorporate multiple conservatisms to account for this source of uncertainty.

In the environment, people are exposed to more than one chemical at a time. However, risks from exposure to multiple chemicals are difficult to predict and most studies that look at toxicity are based on exposure to one or a very small number of compounds. In some cases, exposure to multiple chemicals can cause the body to have a greater (i.e., synergistic) response to a given chemical than it would if the others weren't present. For example, barbiturate drugs have a much greater effect on the nervous system when taken with general anesthetics, alcohol, pain relievers, and other sedative hypnotic drugs (Eaton and Klaassen 2001). In other cases, exposure to multiple agents can reduce toxic effects, for example when one agent blocks the absorption or metabolism of another. For example, ingestion of ipecac or charcoal can prevent the absorption of many toxic substances. When data on the toxicity of a mixture are lacking, most regulatory agencies recommend basing assessments on the toxicity of individual mixture components and assuming that the effects are additive (USEPA 1989; USEPA 2000c; ACGIH 2006). This is the effect most commonly observed when two chemicals are given together (Eaton and Klaassen 2001). However, in any given mixture, a combination of interactive effects is likely; while assuming that the toxicity of each chemical is additive disregards possible synergistic effects, it also disregards antagonistic effects that might decrease toxicity. The appropriateness of assuming effects are additive is supported by experimental evidence suggesting this assumption does not underestimate, and in most cases probably overestimates, risk when chemicals are administered at doses at or near their individual no observed adverse effect levels (Smyth et al. 1970, Ikeda 1988, USEPA 2000c).

For this analysis, drugs from a range of therapeutic classes were considered to provide perspective on possible exposures to and public health impact of pharmaceuticals that share a range of common mechanisms of action (MOAs). The objective here was not to evaluate the cumulative impact of all drugs with a common MOA, but to provide some perspective on whether drugs within certain MOA groups may pose a more significant risk than drugs within other groups and therefore warrant further investigation. In this assessment, the 11 drugs detected in finished or distribution drinking water represented ten distinct modes of action (norfluoxetine is a metabolite of fluoxetine and behaves in a similar manner). Of the compound types that were analyzed, the anticonvulsant/ antipsychotic compounds appear to represent the greatest potential hazard due to their potential carcinogenicity, although in this analysis, these compounds—carbamazepine, phenytoin, risperidone—were all detected at levels below their DWELs. These compounds and compounds with similar biological activity (e.g., with evidence of carcinogenicity in animals) may warrant further investigation. Even though they have some significant toxicity potential, neither of the two statin antilipidemic compounds (atorvastatin and simvastatin) was detected in finished or distribution water.

CHAPTER 10
RISK COMMUNICATION

GENERAL STRATEGIES

Because “safe” levels for most PPCPs and EDCs have not been established through regulation at this time, it is challenging to determine health-based treatment goals. In the absence of the regulatory guidelines and toxicity and exposure data required to set health-based goals, each community must determine whether it will wait for additional information and guidance or take proactive measures to treat and remove trace amounts of PPCPs and EDCs from wastewater and drinking water despite the costs and uncertain benefits. Putting risks in perspective with other common everyday exposures can provide information to assist in the decision-making process. Appendix I provides summary data regarding potential speaking points and communication tools for the dissemination of this report to water industry executives, regulators, the public, and academicians.

Tables 10.1 and 10.2 indicate how many 8-ounce glasses of water per day a person would need to consume, assuming the water contained the maximum-detected finished or distribution water concentration of each target pharmaceutical or EDC, to exceed a daily dose equal to the ADI. For three of the four compounds that were detected, a person would have to drink at least 71,000 glasses of water per day; for atrazine, a person would have to drink 26 glasses of water per day. Table 10.3 compares maximum-detected water concentrations and calculated ADIs and ADI-DWELs for pharmaceuticals to therapeutic dose levels. Daily doses one could get by drinking water with the maximum-detected water concentrations are at least 400,000 times lower than minimum therapeutic doses.

People incur a certain amount of risk every day. Whether they drive a car, cross a street, take medication, drink coffee, or breathe the air, they put themselves at some risk. “Zero risk” does not exist. When considering risks, it is important to note that everyone is at some background level of risk of developing diseases or conditions, such as cancer, during their lifetime. The causes of these conditions are not typically known, but can include genetics, lifestyle factors (e.g., diet, activity levels, general attention to health and well-being), voluntary choices (e.g., smoking, illicit drug use, alcohol consumption), or other “environmental” exposures (e.g., microbes or viruses, chemicals in the workplace or hobby exposure, background radiation). Tobacco smoke, for example, has about 4,000 individual compounds that have been identified as known or probable carcinogens, including benzo(a)pyrene, arsenic, cadmium, polycyclic aromatic hydrocarbons, vinyl chloride, formaldehyde, and nitrosamines (Ropeik and Gray 2002). Smoking is estimated to account for roughly 30 percent of all deaths due to cancer in the United States (Ropeik and Gray 2002). Overall, the probability that a person will develop cancer over the course of a lifetime in the U.S. is slightly less than 1 in 2 for men and slightly more than 1 in 3 for women (ACS 2006).

A 1 in 3 risk can also be expressed as 0.33 or 3.3×10^{-1}. For chemicals that can cause cancer, a risk of developing cancer of 1 in 100,000 (0.00001 or 1×10^{-5}) to 1 in 1,000,000 (0.000001 or 1×10^{-6}) during a lifetime is typically considered an acceptable risk for exposure to a chemical substance (USEPA 1991a, USEPA 2000b). For example, the DWELs for gemfibrozil and phenytoin, derived from SFs calculated using tumor incidence data, are based on an assumed acceptable life-time excess cancer risk of 1 in 1,000,000. If the average U. S. woman were exposed to an additional 1 in 1,000,000 risk (0.000001 or 1×10^{-6}) of developing cancer in her lifetime, her total lifetime cancer risk would rise to 0.330001.

Table 10.1
Daily water consumption rate required to exceed the ADI-DWEL for pharmaceuticals

Drug	Class	ADI-DWEL (µg/L)	Maximum drinking water conc. (µg/L)	Number of glasses of water per day to exceed ADI-DWEL (8-oz glass/d)*
Atenolol	Beta-blocker	70	0.026	23,000
Carbamazepine	Anticonvulsant	12	0.018	5,600
Diazepam	Benzodiazepine tranquilizer	35	0.00033	900,000
Fluoxetine	SSRI antidepressant	34	0.00082	350,000
Norfluoxetine	Fluoxetine metabolite	34	0.00077	370,000
Gemfibrozil	Antilipidemic	45	0.0021	180,000
Meprobamate	Antianxiety agent	260	0.043	51,000
Phenytoin	Anticonvulsant	6.8	0.032	1,800
Risperidone	Antipsychotic	0.49	0.0029	1,400
Sulfamethoxazole	Anti-infective	18,000	0.0030	51,000,000
Triclosan	Antibacterial	2,600	0.0012	18,000,000
Atorvastatin	Antilipidemic	19	<0.00025	>640,000
o-hydroxy atorvastatin	Metabolite	19	<0.00050	>320,000
p-hydroxy atorvastatin	Metabolite	19	<0.00050	>320,000
Diclofenac	NSAID	2,300	<0.00025	>78,000,000
Enalapril	ACE inhibitor	8.1	<0.00025	>270,000
Naproxen	NSAID	20,000	<0.00050	>340,000,000
Simvastatin	Antilipidemic	19	<0.00025	>640,000
Simvastatin hydroxy acid	Metabolite	19	<0.00025	>640,000
Trimethoprim	Antibacterial	6,700	<0.00025	>230,000,000

* 8 oz glasses/day = (DWEL (ug/L) * 2 L/d) * 4.23 8 oz glasses/L/(water concentration (ug/L)

Table 10.2
Daily water consumption rate required to exceed the ADI-DWEL for EDCs

Compound	Use	ADI-DWEL (µg/L)	Maximum drinking water conc. (µg/L)	Number of glasses of water per day to exceed ADI-DWEL (8-oz glass/d)*
Atrazine	Herbicide	180	3	26
Bisphenol A	Industrial chemical	1,800	0.025	610,000
Linuron	Herbicide	70	0.0083	71,000
4-Nonylphenol	Industrial chemical	1,800	0.11	140,000
Butylbenzyl phthalate	Industrial chemical	3,500	<0.050	>590,000
Diethylhexyl phthalate	Industrial chemical	420	<0.10	>36,000
17ß-Estradiol	Hormone	1.8	<0.00050	>30,000
Estrone	Hormone	0.46	<0.00020	>19,000
Ethynylestradiol	Pharmaceutical	0.0035	<0.0010	>30
Lindane	Insecticide	20	<0.010	>17,000
Methoxychlor	Insecticide	0.70	<0.010	>590
Octylphenol	Industrial chemical	5,300	<0.025	>1,800,000
Vinclozolin	Fungicide	420	<0.010	>360,000

* 8 oz glasses/day = [DWEL (ug/L) * 2 L/d) * 4.23 8 oz glasses]/L/water concentration (ug/L)

Table 10.3
Comparison of ADI-DWELs and finished drinking water concentrations with therapeutic doses for pharmaceuticals

Drug	Maximum drinking water conc. (µg/L)	ADI (mg/kg-d)	ADI-DWEL (µg/L)	Minimum therapeutic dose (mg/kg-d)	Minimum therapeutic dose-Equiv water conc (µg/L)
Atenolol	0.026	0.0020	70	0.36	13,000
Carbamazepine	0.018	0.00034	12	1.0	35,000
Diazepam	0.00033	0.0010	35	0.029	1,000
Fluoxetine	0.00082	0.00097	34	0.33	12,000
Norfluoxetine	0.00077	0.00097	34	0.33	12,000
Gemfibrozil	0.0021	0.0013	45	17	600,000
Meprobamate	0.043	0.0075	260	7	250,000
Phenytoin	0.032	0.00019	6.8	10	350,000
Risperidone	0.0029	0.000014	0.49	0.026	910
Sulfamethoxazole	0.0030	0.51	18,000	13	460,000
Triclosan	0.0012	0.075	2,600	NA	NA
Atorvastatin	<0.00025	0.00054	19	0.3	11,000
o-hydroxy atorvastatin	<0.00050	0.00054	19	0.3	11,000
p-hydroxy atorvastatin	<0.00050	0.00054	19	0.3	11,000
Diclofenac	<0.00025	0.067	2,300	1.4	49,000
Enalapril	<0.00025	0.00023	8.1	0.036	1,300
Naproxen	<0.00050	0.570	20,000	4	140,000
Simvastatin	<0.00025	0.00054	19	0.20	7,000
Simvastatin hydroxy acid	<0.00025	0.00054	19	0.20	7,000
Trimethoprim	<0.00025	0.19	6,700	8.0	280,000

Risk estimates are often incorrectly assumed to define a "bright line" between harmful and safe. In general, risk estimates derived using the ADI approach applied here—which is consistent with methodologies used by U.S. EPA and other regulatory agencies—use inputs that, when combined, are intended to represent the upper bound of a possible exposure. For example, assumptions about the concentration of a contaminant to which a person could be exposed are often biased toward the high, either by using data from samples collected at locations where higher-than-average levels are expected (e.g., near an effluent outfall) or by using the maximum or near-maximum measured concentrations. Risk assessors do this because it is essentially impossible to precisely characterize actual exposure concentrations—a limitless number of samples would have to be collected. By biasing the assumed concentration toward the high, risk assessors can be fairly confident that the actual average exposure concentration over time does not exceed the concentration that is assumed.

Similarly, assumptions about the amount, frequency, and duration of exposure are typically biased towards the high. For example, the estimated DWELs assume that a person drinks two liters of tap water (about 8 ½ cups) containing the agent of interest at that concentration every day, year after year after year. However, most people leave home for part of the day to go to work, school, or engage in other activities, and would be expected to drink water or fluids elsewhere, they typically take vacations, and they typically move about to a number of different residences during their lifetime. Also, many people drink other kinds of beverages during the day, including bottled water and purchased juices and sodas, and do not rely on tap water for their entire daily fluid intake.

Risk assessors are also careful to make assumptions about the toxicity of a contaminant that provides a significant margin of safety between the dose that available data suggests is toxic in animals and the dose assumed to be acceptable for a range of human populations. As discussed in Chapter 6, the lowest toxic dose (or the highest nontoxic dose) for the most sensitive endpoint in the most sensitive species is typically selected, and uncertainty factors are applied to account for uncertainties in the available data set. The resulting "acceptable" dose is thus typically at least 300 to 1,000-fold lower than the lowest dose shown to have an adverse effect in toxicity studies, depending on the severity of the toxic effect.

For many PPCPs and EDCs detected in water, exposure of people through water is expected to be small compared to exposures to potentially hazardous compounds through prescription and non-prescription medications, food and beverages, occupational exposures, and residential activities (e.g., cleaning products, personal care products, hobby chemicals, pesticides). When evaluating the results of a risk evaluation, it can be helpful to compare the risk estimates to other more familiar situations, to help people understand the estimates and put the numbers in perspective. Table 10.4 presents lifetime risks associated with other activities or exposures to other environmental agents.

As an example of chemicals that people can be exposed to "naturally" every day, plants produce natural pesticides to defend themselves against fungi, insects, and other animal predators. Although about 10,000 natural pesticides and their break-down products are assumed to occur in foods that make up the human diet, only 71 have been tested adequately for carcinogenicity in rodent bioassays (Gold, Slone, and Ames 2001). Half of these have been shown to be carcinogenic in these tests: these include caffeic acid, d-limonene, and coumarin in such foods and beverages as coffee, lettuce, apples, tomatoes, orange juice, and cinnamon (Gold, Slone, and Ames 2001). Cooking and preparation of food also produce chemicals that are carcinogenic in rodent studies. Furfural is a chemical formed naturally when sugars are heated, and is a widespread constituent of food flavor. A variety of mutagenic and carcinogenic heterocyclic amines, such as benzo(a)pyrene, are formed when meat, chicken or fish is cooked, particularly when charred (Gold, Slone, and Ames 2001). Epidemiological studies indicate that all types of alcoholic beverages are associated with increased cancer risk, suggesting that ethyl alcohol produced by fermentation is the carcinogenic component rather than any particular type of beverage (Gold, Slone, and Ames 2001).

Table 10.4
Comparison of lifetime risks associated with different events or situations

Event or situation	Approximate lifetime risk
Risks of accidental death	
Dying in a motor vehicle accident	1 in 100 *
Dying in a motor vehicle accident per 100 miles traveled	1 in 1,000,000 †
Dying while a pedestrian	1 in 1,000 *
Drowning	1 in 1,000 ‡
Dying from a fall at home	3 in 1,000 *
Dying from a lightening strike	1 in 100,000 ‡
Dying while riding a bicycle	5 in 1,000,000 §
Risks of developing cancer from environmental agents	
Skin cancer, primarily from exposure to the sun	1 in 5 **
Cigarette smoking (a pack or more per day)	8 in 100 ††
Outside radiation (radon and cosmic rays)	1 in 1,000 ††
Environmental tobacco smoke (death from cancer)	1 in 1,000 ‡‡
Human-made chemicals in indoor air at home	2 in 10,000 ††
Outdoor air in industrialized area	1 in 10,000 ††
Human-made chemicals in drinking water (chlorination)	1 in 100,000 ††
Consumption of 2 oz. of peanut butter per week (naturally occurring aflatoxin)	8 in 100,000 ††
Consumption of one meal per year of small Lake Michigan trout	1 in 100,000 ††

* Source: NSC 2000
† Source: U.S. DOT 2000
‡ Source: NSC 2006
§ Source: IIHS 2005
** Source: AAD 2006
†† Source: USEPA 1991b
‡‡ Source: USEPA 1993b

One way to put possible cancer risks from exposure to pharmaceuticals in drinking water in perspective is to compare the relative carcinogenic risk of daily exposure to these agents with other commonly present agents. For example, rodent testing suggests that gemfibrozil and phenytoin, two pharmaceuticals that have been detected in finished drinking water water, are carcinogenic. Risks from daily exposure to these agents in water at their maximum detected concentrations (assuming consumption of 2 liters of this water per day) can be compared to agents present in different food products or in drugs taken voluntarily for therapeutic purposes, using information on the carcinogenic potency of nearly 1,500 chemicals based on data from long-term animal testing that has been compiled in a unique database by the Carcinogenic Potency Project at the Lawrence Berkeley Laboratory of the University of California (Gold et al. 2006). "TD50" values (i.e., chronic dose-rates in mg/kg body weight/day that would induce tumors in half the test animals at the end of a standard lifespan) for these compounds can be compared, incorporating assumptions about average daily human exposure rates to each chemical. Table 10.5 shows these comparisons for gemfibrozil and phenytoin at the maximum detected finished water concentrations versus other chemicals to which people are typically exposed, including "naturally" present compounds in foods. These comparisons show that carcinogenic risks of exposure to gemfibrozil and phenytoin at measured levels in drinking water are on the order of more than 1,000,000 times lower than carcinogenic risks from daily ingestion of alcoholic beverages and consumption of other compounds naturally present in foods.

Table 10.5
Estimated cancer risks from exposure to pharmaceuticals in water relative to other commonly present carcinogens based on the human exposure/ rodent potency index (HERP)

Chemical	Description	Chronic Human Exposure (µg/day)*	Rodent Potency Index (TD_{50}) (mg/kg/day)†	Human Exposure/ Rodent Potency Index (%)‡	Fraction of Highest Food-Related HERP
Compounds detected in finished drinking water					
Gemfibrozil	Based on maximum finished water concentration (0.0021 µg/L)	0.0042	125	<0.0000001	<0.000000028
Phenytoin	Based on maximum finished water concentration (0.032 µg/L)	0.064	59.1	<0.00001	<0.0000028
Compounds naturally present in food					
Aflatoxin (fungal toxin)	In: Peanut butter	0. 018	0.0032	0.008	0.0022
Caffeic acid (natural pesticide)	In: Coffee (~0.5 oz.)	20,800	297	0.1	0.028
	Lettuce (14.9 g)	7,900		0.04	0.011
	Tomato (89 g)	5,460		0.03	0.0083
	Apple (32.0 g)	3,400		0.02	0.0055
Coumarin (natural pesticide)	In: Cinnamon (21.9 mg)	65	13.9	0.007	0.0019
d-Limonene (natural pesticide)	In: Orange juice (~4.5 oz.)	4,280	204	0.03	0.0083

(continued)

Table 10.5 (Continued)

Chemical	Description	Chronic Human Exposure (μg/day)*	Rodent Potency Index (TD_{50}) (mg/kg/day)†	Human Exposure/ Rodent Potency Index (%)‡	Fraction of Highest Food-Related HERP
Compounds naturally present in food (continued)					
Ethyl alcohol (from fermentation)	In alcoholic drinks: All types Beer (~8 oz.) Wine (~1 oz.)	22,800,000 11,700,000 3,670,000	9,110	3.6 1.8 0.6	1 0.5 0.17
Furfural (formed when sugars are heated)	In: Bread (~3 slices)	867	197	0.004	0.0011
PhIP (a pyrene formed during cooking)	In: Hamburger, pan fried (one 3 oz. serving)	0. 176	1.64	0.0002	0.000056
Compounds commonly present in the environment					
Formaldehyde	In conventional home air (14 hours/day)	598	2.19	0.4	0.11
Pharmaceuticals					
Clofibrate	Cholesterol lowering agent	2,000,000	169	17	4.7
Fluvastatin	Cholesterol lowering agent; one of the top 200 most prescribed drugs in 2003	20,000	125	0.2	0.056
Gemfibrozil	Cholesterol lowering agent; 5,797,000 U.S. prescriptions dispensed in 2005	1,200,000	247	6.9	
Lovastatin	Cholesterol lowering agent; 10,328,000 U.S. prescriptions dispensed in 2005	20,000	515	0.06	

Source: Gold et al. 2006

* Average U.S. exposure or consumption data for which human exposure could be chronic for a lifetime. For pharmaceuticals the doses are recommended doses.

† A TD50 value is the tumorigenic dose-rate for rodents, defined as the chronic dose-rate in mg/kg body weight/day that would induce tumors in half the test animals at the end of a standard lifespan for a species.

‡ Human Exposure/Rodent Potency index (HERP) -- indicates the percentage of the rodent TD50 (in mg/kg/day) a person receives from average daily exposure to a chemical over a lifetime (mg/kg/day).

CHAPTER 11
CONCLUSIONS AND RECOMMENDATIONS

DISCUSSION

By applying highly sensitive and selective analytical methods, it is now possible to detect and quantify a vast number of trace chemicals in water at extremely minute levels. Even the most advanced water treatment technologies are incapable of reducing the concentrations of all contaminants to less than the detection limit of modern analytical techniques. Over the past decade alone, methods have been published that are capable of detecting emerging contaminants in water in parts per quadrillion levels. For instance, Huang and Sedlak published methodology for analyzing steroid hormones in surface water with a reporting limit of 50 parts per quadrillion (0.00000000005 g/L) (Huang and Sedlak 2001b), which can be thought of as less than one minute in 31.7 million years of history. Today, many analytical methods are capable of low to sub-ng/L reporting limits for trace contaminants in water (Ternes 2001; Loos et al. 2003; Batt and Aga 2005; Trenholm et al. 2006; Vanderford and Snyder 2006). Through application of these sensitive methods, it is apparent that trace contaminants are present in many waters. The continued drive towards costly monitoring programs is of little value without determination of threshold concentrations that are protective of human health.

Pharmaceuticals and steroid hormones have a unique connotation in the minds of the public since these compounds primarily enter the environment through human excretion and are generally processed through wastewater treatment plants. The psychological perception linking human discharge and tap water contamination is difficult to debate. However, a clear explanation of concentration in terms of relevance to human health and relative risks provide critical data to better perceive the infinitesimally minute levels at which these compounds have been detected. One fact that draws interest is that the U.S. government has studied these contaminants for more than 30 years (Tabak and Bunch 1970, Garrison et al. 1975). These early reports indicate that concentrations of steroid hormones and certain pharmaceuticals have dropped off dramatically, most likely due to increased efficiency of wastewater treatment (i.e., greater efforts towards nutrient removal). Conversely, new chemicals for a diversity of applications are continually introduced to the market, essentially all of which will be detectable at some concentration in water. The fact is, all chemicals in commerce can and will be present in water at some measurable concentration, perhaps not measurable today, but almost certainly measurable with tomorrow's analytical technology.

One may imagine that a simple solution to emerging contaminant threats is simply to add additional treatment barriers. We know that advanced processes such as ozone, activated carbon, UV-advanced oxidation, and reverse osmosis can greatly reduce the concentrations of most (but not all) trace contaminants. However, today's growing concerns regarding climate change, carbon footprint, and economic instability all warrant careful evaluation of cost-benefit before moving towards additional treatment that may not significantly improve human. Current advanced treatment goals for emerging contaminants are generally based upon achievement of non-detection using non-standardized analytical methods; however, we must consider that not long ago, the vast majority of today's emerging contaminants would have been undetectable. Does this mean that we were safer then than now? The fundamental point of departure in our actions today must be protection of health, not simply protection against perception. Since perception is driven mostly by presence and absence, even large capital expenditures for advanced treatment today to reach non-detect will likely fail against tomorrow's more sensitive techniques. A key point is that today's non-detects will likely be tomorrow's emerging contaminants.

To this end, the project presented here set out to scour the available literature for information regarding the human health toxicity of a representative group of pharmaceutical and potential EDC indicator compounds. These compounds were largely chosen as those with the greatest potential for toxicity and occurrence in drinking water. In order to conduct a hazard assessment, robust analytical methods were developed and employed to screen water from twenty sites across the U.S. Based upon the toxicology review and occurrence data, a risk assessment based on standard regulatory procedures was conducted to determine if these indicator compounds occurred in U.S. drinking waters at levels that may pose risk to human health.

CONCLUSIONS

An indicator suite of 62 chemicals was monitored in U.S. drinking waters. These included 20 pharmaceuticals, 26 potential EDCs, five steroid hormones, and 11 phytoestrogens. Robust analytical methodology was developed specifically for this study and subsequently published in peer-reviewed journals (Vanderford and Snyder 2006, Trenholm et al. 2006). Samples from geographically diverse utilities across the U.S. were monitored and only a few of the target analytes were frequently detected (>50% frequency) in U.S. drinking waters. These compounds were the herbicide atrazine (potential EDC), meprobamate (anti-anxiety pharmaceutical), and phenytoin (anti-epileptic pharmaceutical). For those compounds detected in any finished drinking water sample, Table 11.1 provides a summary which includes the maximum concentration detected, the drinking water equivalent level (DWEL), and the recommended method reporting limit. Data in Table 11.1 are presented in µg/L as opposed to ng/L shown elsewhere, in order to accommodate the proper number of significant figures in the DWEL. The maximum concentration is the highest value recorded in any finished water sample tested in this study. Where the maximum value was less than one ng/L, the value was increased to one ng/L in order to preserve significant figures and present a conservative approach. The DWEL is reflective of the human health risk assessment and based upon the acceptable daily intake (which includes conservative uncertainty factors), a body mass of 70 kg, and a water consumption rate of 2 liters/day. Drinking water regulations are often calculated in DWELs as it is directly comparable to occurrence concentrations. The third column in Table 11.1 provides the number of liters a person would need to consume in one day in order to reach the acceptable daily intake. In other words, this volume of water could be consumed per day without exceeding the dose protective of human health. The final column in Table 11.1 provides a recommended method reporting limit (MRL). The recommended MRL is approximately 100-fold lower than the DWEL and would provide two-logs of detection below the concentration calculated to have no negative impact to human health. By using MRLs based on health effects, with two or more orders of magnitude "safety factors", meaningful occurrence data and treatment goals can be established. Interestingly, using the recommended MRLs in Table 11.1, only risperidone and atrazine would have come close to the MRLs. Risperidone was detected one time in a distribution sample and never in the finished water. The MRL for risperidone was 2.5 ng/L using the analytical method developed in this project, while this single documented occurrence in drinking water was at 2.9 ng/L. Since this compound was only detected in one drinking water sample and very close to the limits of detection, the occurrence of risperidone in U.S. drinking waters is questionable. On the other hand, atrazine was detected in up to 87% of the U.S. drinking waters tested with a maximum concentration of 1.0 μg/L.

Table 11.1
Summary of Drinking Water Occurrence, Risk, and Recommended Method Reporting Limits

	Max Drinking Water Conc (μg/L)	ADI as a DWEL (μg/L)	No of Liters per day to meet ADI dose*	Recommended MRL (μg/L)
Bisphenol A	0.025	1,800	144,000	20
Risperidone	0.003	0.49	330	0.005
Phenytoin	0.032	6.8	430	0.1
Carbamazepine	0.018	12	1,300	0.1
Diazepam	0.001	35	70,000	0.4
Fluoxetine	0.001	34	68,000	0.3
Norfluoxetine	0.001	34	68,000	0.3
Gemfibrozil	0.002	45	45,000	0.5
Atenolol	0.026	70	5,400	0.7
Atrazine	1.0	3	6	0.03
Meprobamate	0.043	260	12,000	3.0
Linuron	0.008	70	18,000	0.7
Triclosan	0.001	2,600	5,200,000	30
4-Nonylphenol	0.11	1,800	33,000	20
Sulfamethoxazole	0.003	18,000	1,200,000	200

* Assumes an average weight of 70 kg and 2 L/day consumption.

The E-screen assay is capable of nearly unfathomable sensitivity, with a calculated detection limit of approximately 0.03 ng/L in water. This level is even lower than the majority of blank (control) samples under the most pristine of laboratory conditions. Based upon a statistical evaluation of blanks and E-screen sensitivity, the reporting limit for this study was 0.16 ng/L (160 parts per quadrillion) of estradiol equivalents. Only three drinking water samples had estrogenicity above the E-screen MRL. However, these "detections" were all at less than one part per trillion. In comparison, common food/beverage items frequently contained estrogenicity values orders of magnitude greater than the rare detections in water. The point here is not that the foods/beverages evaluated pose risk, but rather that the E-screen is an inappropriate metric for human health protection. Significantly more research is needed in order to use a cellular bioassay alone to protect public health from the insult of trace contaminants.

In summary, this study found that while some trace pharmaceuticals and potential EDCs can be detected in U.S. drinking waters, the levels at which they occur are far below any human health relevance. In the calculation of the levels below which no significant risk to human health occurs, we have used conservative uncertainty factors to protect the most susceptible of our population. These conclusions are based upon peer-reviewed published literature and occurrence data from robust analytical methodology. More research is certainly warranted, but from the data available for the calculations used in this report, the fears over human health adversities from trace pharmaceuticals and potential EDCs are largely unsubstantiated. The fact that pharmaceuticals and EDCs are mainstream issues is not due to the deterioration of water quality in the U.S., but rather dramatic improvements to analytical capabilities in recent decades. From the early reports by the U.S. federal government more than 30 years ago demonstrating the presence of pharmaceuticals and steroid hormones in source waters at parts-per-million levels, to the recent media coverage regarding these chemicals in tap water in the range of parts per trillion, the only significant change has been the availability of methodology to identify contaminants minuscule levels in water. A noteworthy observation is that, had this research team not had the ability to evaluate compounds at the sub-ppb level, none of the compounds discussed in this study would have been found. It is true that new pharmaceuticals continue to enter the market each year, but is also true that water treatment continues to improve and evolve. From regulations to protect the public from pathogens of genuine health concern in drinking water to total maximum daily load regulations of nutrients from wastewater, the U.S. water industry continues to advance to remain leaders in the protection of public and environmental health.

RECOMMENDATIONS FOR UTILITIES

The multitude of projects undertaken by water utilities and agencies on topics that are not mandated under any regulatory requirements demonstrates the proactive nature and concern for public welfare. Utilities are encouraged to continue working together with the research foundation and other partners to remain abreast of the latest developments in water quality, treatment, and public communication. In light of recent media coverage, an increasing number of U.S. water utilities desire to establish monitoring programs for emerging contaminants, especially pharmaceuticals. The project team commends these efforts, but wishes to convey a sense of caution and recommendations for these programs.

It is clear that quality assurance and quality control (QA/QC) are critical to trace contaminant analysis. Methods capable of parts per trillion (and lower) detection and quantitation are available; however, increased rigor is needed in order to have confidence in the results. Without a solid QA/QC plan, which goes beyond the conventional metrics used for standard methods, misinformation is not only possible, but often probable. Even the most widely cited study on pharmaceuticals in U.S. waters (Kolpin et al. 2002) faced analytical challenges. According to the information provided on the USGS website (http://toxics.usgs.gov/regional/est_errata.html), some steroid hormone data from Kolpin 2002 were later "rejected as having potential interferences". Another recent example includes claims from a major newspaper that acetaminophen (commonly known by the trade name Tylenol™) occurred in a major U.S. city's drinking water at low ng/L levels. This finding is quite probably an analytical artifact. First, acetaminophen can frequently occur in blank (control) samples as false positives. The USGS found acetaminophen in 59.1% of their blanks with an average value of 16 ng/L, a value greater than what was purported by the media in some U.S. drinking waters. Moreover, as demonstrated in AwwaRF project 2758, acetaminophen is rapidly oxidized by either free chlorine or chloramine, which is present in the tap water of all major U.S. cities. In a method published in 2003 by this project team, the challenges of inaccurate quantitation due to matrix inferences was also addressed (Vanderford et al. 2003). The take-home message is that utilities should enter into unregulated contaminant monitoring programs carefully. It is suggested that a rigorous QA/QC program be developed that includes frequent field and laboratory blanks, random replication, blinding sample names from the laboratory, and random matrix spikes near the MRLs. It is also suggested that utilities insist on MRLs based upon human health protection when possible, as opposed to seeking the lower possible reporting limits based simply on analytical capability.

We encourage water utilities to continue open dialog with the public regarding emerging water quality issues. Appendix I provides a series of speaking/discussion points that have been developed through our discussions with state regulators and the participating utilities. We strongly recommend that utilities review the materials in Appendix I for valuable information regarding the communication of the findings within this report, and the topic of emerging contaminants in general.

Within this report, we provide tools that are of potential benefit to utilities; however, each utility must develop speaking points that address their specific communities. We recommend that utilities avoid rushing into major capital investments based solely upon the current state of knowledge on pharmaceuticals and potential EDCs; however, when replacing or expanding infrastructure the ancillary benefits of particular treatment options for reducing trace contaminants should be considered. We encourage a holistic view of water and energy sustainability in concert with concerns regarding emerging contaminants. This project investigated a limited number of pharmaceuticals and potential EDCs, while the number of possible chemicals in these classes is in the thousands. Utilities should expect that additional trace contaminants will continue to be unearthed as analytical capabilities become more sensitive and more readily available at reduced costs. The determination of toxicological relevance of emerging contaminants is of pinnacle importance. Just as detection alone does not infer harm, non-detection does not ensure safety. By using toxicology to guide reasonable analytical detection limits and meaningful treatment goals, utilities can maximize their limited resources for the best protection of human health.

REFERENCES

AAD (American Academy of Dermatology). 2006. 2007 *Skin Cancer Fact Sheet.* [Online] Available: <www.aad.org/SkinCancerNews/WhatIsSkinCancer/SCancerFacts.html> [cited on June 1, 2007]

ACGIH, A. C. o. G. I. H. 2006. 2006 TLVs and BEIs Based on the Documentation of the Threshold Limit Values for Chemical Substances and Physical Agents & Biological Exposure Indices. Cincinnati, OH.

ACS, A.C.S. 2006. Cancer facts & figures 2006. Atlanta, GA.

ATSDR (Agency for Toxic Substances and Disease Registry) 2007. Minimal Risk Levels (MRLs) for Hazardous Substances. [Online]. Available: <www.atsdr.cdc.gov/mrls.html.>. [cited June 1, 2007]

CADHS, C. D. o. H. S. 2006. Drinking Water Notification Levels and Response Levels: An Overview. Drinking Water Program. Retrieved June 28, 2006, from: www.dhs.ca.gov/ps/ddwem/chemicals/AL/notificationoverview.pdf.

CalEPA (California Environmental Protection Agency) 2006. Chemicals Known to the State to Cause Cancer or Reproductive Toxicity. [Online]. Available: <www.oehha.ca.gov/prop65/prop65_list/files/P65single120806.pdf>. [cited May 22, 2006]

EC-BKH, E.C.-B. 2000. Towards the establishment of a priority list of substances for further evaluation of their role in endocrine disruption: preparation of a candidate list of substances as a basis for priority setting, In *Final report*. Edited by: European Commission- BKH.

EC, E.C. 2002. Study on the scientific evaluation of 12 substances in the context of endocrine disrupter priority list of actions. Edited by Buckinghamshire, United Kingdom: European Commission.

EDSP, E.D.S.P. 2003. Reference chemicals used in EDSP tier 1 prevalidation studies, In *Prepared for the Endocrine Disrupter Methods Validation Subcommittee*. Edited by Washington, D.C.: United States Environmental Protection Agency, Endocrine Disrupter Screening Program.

EDSTAC, E.D.S.A.T.A.C. 1998. Endocrine Disruptor Screening and Testing Advisory Committee (EDSTAC) Final Report. Edited by E. D. S. a. T. A. C. United States Environmental Protection Agency, ed. Washington, D.C.: Endocrine Disruptor Screening and Testing Advisory Committee (EDSTAC).

FDA-EDKB, U.S.F.A.D.A.-E.D.K.B. 2007. Endocrine Disruptor Knowledge Base (EDKB). Edited by Rockville, MD: United States Food and Drug Administration.

FDA (Food and Drug Administration) 1995. Guideline for Industry: Dose Selection for Carcinogenicity Studies of Pharmaceuticals. [Online]. Available: <www.fda.gov/Cder/guidance/ichs1c.pdf>.

FDA (Food and Drug Administration) 2000. Professional Drug Information Database listing for Anticonvulsants, Hydantoin (Systemic). [Online]. Available: <www.drugs.com/MMX/phenytoin.html>. [cited June 1, 2007]

FDA (Food and Drug Administration), F.A.D.A. 2006. Questions and Answers: Over-the-Counter Drug Products Public Hearing.

FDA (Food and Drug Administration) 2007. Adverse Event Reporting System (AERS). [Online]. Available: <www.fda.gov/cder/aers/default.htm>.

GWRC, G.W.R.C. 2003. Endocrine disrupting compounds: Priority list of EDCs. Edited by London, United Kingdom: Global Water Research Coalition.

IARC (International Agency for Research on Cancer) 1996a. Clofibrate. IARC Monogr Eval Carcinog Risks Hum [Online]. Available: <www.inchem.org/documents/iarc/vol66/clofibrate.html>.

IARC (International Agency for Research on Cancer) 1996b. Gemfibrozil. IARC Monogr Eval Carcinog Risks Hum [Online]. Available: <www.inchem.org/documents/iarc/vol66/gemfibrozil.html>.

IARC (International Agency for Research on Cancer) 1996c. Phenytoin. IARC Monogr Eval Carcinog Risks Hum. [Online]. Available: <www.inchem.org/documents/iarc/vol66/phenytoin.html>.

IEH, I.F.E.A.H. 2005. Chemicals purported to be endocrine disruptors: a compilation of published lists (web report W20). Edited by Leicester, United Kingdom: Medical Research Council, Institute for Environment and Health.

IIHS (Insurance Institute for Highway Safety). 2005. *Fatality Facts: Bicycles*. Highway Loss Data Institute. [Online]. Available: <www.iihs.org/research/fatality_facts/bicycles.html >[cited on June 1, 2007]

JECFA (Joint FAO/WHO Expert Committee on Food Additives). 2000. Toxicological Evaluation of Certain Veterinary Drug Residues in Food. WHO Food Additive Series: 43. Prepared by the fifty-second meeting of the Joint FAO/WHO Expert Committee on Food Additives (JECFA). Geneva, Switzerland: World Health Organization, International Programme on Chemical Safety. [Online]. Available: <http://www.inchem.org/documents/jecfa/jecmono/v43jec05.htm>.

NDCHealth 2005. Micromedex Physicians' Desk Reference [Online].

NRC, N.R.C. 1999. Hormonally active agents in the environment. Edited by Washington, D.C.: National Academies Press.

NRC, N.R.C. 2005. Health implications of perchlorate ingestion. Edited by Washington, D.C.: National Academies Press.

NSC (National Safety Council). 2000. *Report on Injuries in America: National Safety Council 1999 Injury Facts*. [Online] Revised June 6, 2000. Available: www.nsc.org/lrs/statinfo/99report.htm [cited June 1, 2007]

NSC (National Safety Council). 2006. What Are the Odds of Dying? [Online] Revised August 3, 2006. Available: <www.nsc.org/lrs/statinfo/odds.htm> [cited June 1, 2007]

NTP-CERHR, N.T.P.-C.F.T.E.O.R.T.H.R. 2003. NTP-CERHR monograph on the potential human reproductive and developmental effects of Butyl Benzyl Phthalate (BBP), In *NIH publication no. 03-4487*. Edited by Research Triangle Park, NC: United States Department of Health and Human Services, National Toxicology Program (NTP), Center for the Evaluation of Risks to Human Reproduction (CERHR).

NTP-CERHR, N.T.P.-C.F.T.E.O.R.T.H.R. 2006. NTP-CERHR monograph on the potential human reproductive and developmental effects of Di(2-Ethyhexyl) Phthalate (DEHP), In *NIH publication no. 06-4476*. Edited by Research Triangle Park, NC: United States Department of Health and Human Services, National toxicology Program (NTP), Center for the Evaluation of Risks to Human Reproduction (CERHR).

NTP-CERHR, N.T.P.-C.F.T.E.O.R.T.H.R. 2007. NTP-CERHR expert panel report on the reproductive and developmental toxicity of Bisphenol A, In *NTP-CERHR-BPA-07*. Edited by Research Triangle Park, NC: United States Department of Health and Human Services, National Toxicology Program (NTP), Center for the Evaluation of Risks to Human Reproduction.

NTP, N.T.P. 1993. Toxicology and Carcinogenesis Studies of 5,5-Diphenylhydantoin (CAS No. 57-41-0) (Phenytoin) in F344/N Rats and B6C3F1 Mice (Feed Studies). *Natl Toxicol Program Tech Rep Ser*, 404:1-303.

NTP (National Toxicology Program) 2000. NTP technical report on the toxicity studies of Carisoprodol (CAS No. 78-44-4) administered by Gavage to F344/N rats and B6CF1 mice [Online]. Available: <http://ntp.niehs.nih.gov/ntp/htdocs/ST_rpts/tox056.pdf>. [cited June 1, 2007]

NTP, N.T.P. 2003. Pubertal toxicity study of Vinclozolin, Flutamide and Phenobarbital in male Sprague Dawley rats and Methoxychlor, Ethinyl Estradiol and Phenobarbital in female Sprague Dawley rats when administered in corn oil by oral gavage. Edited by Research Triangle Park, NC: National Toxicology Program (NTP).

NTP (National Toxicology Program) 2004. NTP-CERHR monograph on the potential human reproductive and developmental effects of Fluoxetine [Online]. Available: <http://cerhr.niehs.nih.gov/chemicals/fluoxetine/fluoxetine_monograph.pdf>. [cited May 29, 2007]

US DOT (U.S. Department of Transportation). 2000. North American Transportation in Figures. U.S. Department of Transportation, Bureau of Transportation Statistics. BTS00-05, Washington, DC. Online. Available: <www.bts.gov/itt/natf.html>

USEPA-IRIS. 2005. Perchlorate and Perchlorate salts, In *Integrated Risk Information System (IRIS) summary*. Edited by Washington, DC: United States Environmental Protection Agency.

USEPA-OPPTS. 2006a. Triazine cumulative risk assessment and Atrazine, Simazine, and Propazine decisions. Edited by Washington, DC: Environmental Protection Agency-Office of Pollution Prevention and Toxics.

USEPA (United States Environmental Protection Agency). 1989. Risk Assessment Guidance for Superfund: Volume 1-Human Health Evaluation Manual, Part A, Interim Final (9285.701A). Office of Solid Waste and Emergency Response: Washington, D.C.

USEPA. 1991a. Role of the Baseline Risk Assessment in Superfund Remedy Selection Decisions. *Office of Solid Waste and Emergency Response*:OSWER Directive 9355.9350-9330.

USEPA. 1991b. *Environmental Risk: Your Guide to Analyzing and Reducing Risk*. Region 5. EPA/905/9-91/017. October.

USEPA 1993a. Reference Dose (RfD): Description and Use in Health Risk Assessments [Online]. Available: <www.epa.gov/iris/rfd.htm>.

USEPA. 1993b. Respiratory Health Effects of Passive Smoking: Lung Cancer and Disorders. Office of Research and Development. EPA/600/6-90/006F.

USEPA. 2000. Vinclozolin; pesticide tolerances. Federal Register 65(138):44453-44468. July 18, 2000. [Online]. Available: <http://www.epa.gov/EPA-PEST/2000/July/Day-18/p18099.htm>.

USEPA 2000a. Methodology for Deriving Ambient Water Quality Criteria for the Protection of Human Health [Online]. Available: <www.epa.gov/waterscience/criteria/humanhealth/method/complete.pdf#search=%22Methodology%20for%20Deriving%20Ambient%20Water%20Quality%20Criteria%20for%20the%20Protection%20of%22>.

USEPA 2000b. Strategy for Research on Environmental Risks to Children [Online]. Available: <www.epa.gov/ord/htm/researchstrategies.htm#rs04>.

USEPA. 2000c. Supplementary Guidance for Conducting Health Risk Assessment of Chemical Mixtures (EPA/630/R-00/002). *Risk Assessment Forum*:Washington, D.C. External Scientific Peer Review Draft.

USEPA. 2001. Exploration of Aging and Toxic Response [Online]. Available: <http://oaspub.epa.gov/eims/eimsapi.dispdetail?deid=29411>.

USEPA. 2002. A Review of the Reference Dose and Reference Concentration Processes [Online]. Available: <www.epa.gov/iris/RFD_FINAL[1].pdf>.

USEPA. 2005. Guidelines for Carcinogen Risk Assessment [Online]. Available: <www.epa.gov/iris/backgr-d.htm>.

USEPA. 2006. Setting Standards for Safe Drinking Water [Online]. Available: <www.epa.gov/ogwdw/standard/setting.html>.

USEPA. 2007a. Glossary of IRIS Terms [Online]. Available: <www.epa.gov/iris/gloss8.htm>. [cited June 1, 2007]

USEPA. 2007b. Drinking water contaminants: list of contaminants and their maximum contaminant levels (MCLs). Edited by Washington, D.C.: U.S. Environmental Protection Agency, Office of Ground Water and Drinking Water.

WHO-IPCS, W.H.O.-I.P.O.C.S. 2002a. Global assessment of the state-of-the-science of endocrine disruptors, In *World Health Organization technical report series*. Edited by Geneva, Switzerland: World Health Organization.

WHO-IPCS, W.H.O.-I.P.O.C.S. 2002b. Global assessment of the State-of-the-science of endocrine disruptors, In *WHO/PCS/EDC/02.2*. Edited by T. DamstraS. BarlowA. BergmanR. Kavlock and G. Van Der Kraak, eds. Geneva, Switzerland: World Health Organization, International Programme on Chemical Safety.

WHO, W.H.O. 1994. World Health Organization Environmental Health Criteria 170. Assessing human health risks of chemicals: Derivation of guidance values for health-based exposure limits [Online]. Available: <www.inchem.org/documents/ehc/ehc/ehc170.htm#SectionNumber:2.2>.

Adams, C. W., Y. Loftin, K. Meyer, M. 2002. Removal of Antibiotics from Surface and Distilled Water in Conventional Water Treatment Processes. *Journal of Environmental Engineering*. 128(3): 253.

Adams, N.R. 1998. Natural and anthropogenic environmental oestrogens: the scientific basis for risk assessment. Clover phyto-oestrogens in sheep in Western Australia. *Pure & Appl. Chem.*, 70:1855-1862.

Adlercreutz, H. 1984. Does fiber-rich food containing animal lignin precursors protect against both colon and breast cancer? An extension of the "fiber hypothesis.". *Gastroenterology*, 86:761-764.

Adlercreutz, H., H. Honjo, A. Higashi, T. Fotsis, E. Hamalainen, T. Hasegawa, and H. Okada. 1991. Urinary excretion of lignans and isoflavonoid phytoestrogens in Japanese men and women consuming traditional Japanese diet. *Am. J. Clin. Nutr.*, 54:1093-1100.

Aerni, H.-R., B. Kobler, B. Rutishauser, F. Wettstein, R. Fischer, W. Giger, A. Hungerbühler, M.D. Marazuela, A. Peter, R. Schönenberger, A.C. Vögeli, M.F. Suter, and R.L. Eggen. 2004. Combined biological and chemical assessment of estrogenic activities in wastewater treatment plant effluents. *Analytical and Bioanalytical Chemistry*, 378(3):688-696.

Aherne, G.W., and R. Briggs. 1989. The Relevance of the Presence of Certain Synthetic Steroids in the Aquatic Environment. *J. Pharm. Pharmacol.*, 41:735-736.

Andreolini, F., C. Borra, F. Caccamo, A. Di Corcia, and R. Samperi. 1987. Estrogen conjugates in late-pregnancy fluids: extraction and group separation by a graphitized carbon black cartridge and quantification by high-performance liquid chromatography. *Analytical Chemistry*, 59(13):1720-1725.

Antignac, J.-P., R. Cariou, B. Le Bizec, J.-P. Cravedi, and F. Andre. 2003. Identification of phytoestrogens in bovine milk using liquid chromatography/electrospray tandem mass spectrometry. *Rapid Communications In Mass Spectrometry*, 17(12):1256-1264.

Ash, R.J., B. Mauck, and M. Morgan. 2002. Antibiotic Resistance of Gram-Negative Bacteria in Rivers, United States. *Emerg Infect Dis.*, 8(7):713-716.

Ashby, J., H. Tinwell, et al. 1997. Normal sexual development of rats exposed to butyl benzyl phthalate from conception to weaning. *Regul Toxicol Pharmacol.* 26(1 Pt 1): 102-18.

Atkinson, S., M.J. Atkinson, and A.M. Tarrant. 2003. Estrogens from Sewage in Coastal Marine Environments. *Environmental Health Perspectives*, 111(4):531.

Ayscough, N.J., J.Fawell, G.Franklin, and W.Young. 2000. Review of Human Pharmaceuticals in the Environment. R&D Tech. Report P390. WRc Report No. EA4761. WRc-NSF Ltd, Buckinghamshire. Edited by: Bristol: Environment Agency.

Balch, G., and C. Metcalfe. 2006. Developmental effects in Japanese medaka (Oryzias latipes) exposed to nonylphenol ethoxylates and their degradation products. *Chemosphere*, 62(8):1214-1223.

Barbolt, T.A. 2002. Chemistry and safety of Triclosan, and its use as an antimicrobial coating on coated VICRYL* Plus antibacterial suture (coated Polyglactin 910 suture with Triclosan). *Surgical Infections*, 3(Suppl):S-45-53.

Barcelo, D., C. Postigo, and M.J.L.D. Alda. 2008. Fully Automated Determination in the Low nanogram per Liter Level of Different Classes of Drugs of Abuse in Sewage Water by On-Line Solid-Phae Extraction Liquid Chromatography Electrospray Tandem Mass Spectrometry. *Analytical Chemistry*, 80(9):3123-3134.

Batt, A. L. and Aga, D. S. 2005. Simultaneous analysis of multiple classes of antibiotics by ion trap LC/MS/MS for assessing surface water and groundwater contamination. *Analytical Chemistry*, **77**(9): 2940-2947.

Belzer R.B., Bruce G.M., Peterson M.K. and Pleus R.C. 2004. Using Comparative Exposure Analysis to Validate Low-Dose Human Health Risk Assessment: The Case of Perchlorate. In Linkov, I. and Ramadan, A., Eds. *Comparative Risk Assessment and Environmental Decision Making*. Kluwer. pp. 57-74.

Bennetts, H.W., E.J. Underwood, and F.L. Shier. 1946. A specific breeding problem of sheep on subterranean clover pastures in western Australia. *Aust. Vet. J.*, 22:2-12.

Bevans, H.E., S.L. Goodbred, J.F. Miesner, S.A. Watkins, T.S. Gross, N.D. Denslow, and T. Schoeb. 1996. Synthetic organic compounds and carp endocrinology and histology in Las Vegas Wash and Las Vegas and Callville Bays of Lake Mead, Nevada, 1992 and 1995. *Water-Resources Investigations Report 96-4266.*

Bittigau, P., M. Sifringer, K. Genz, E. Reith, D. Pospischil, S. Govindarajalu, M. Dzietko, S. Pesditschek, I. Mai, K. Dikranian, J.W. Olney, and C. Ikonomidou. 2002. Antiepileptic drugs and apoptotic neurodegeneration in the developing brain. *Proc Natl Acad Sci USA*, 99(23):15089-15094.

Borgert C.J., LaKind J.S., Witorsch R.J. 2003. A Critical Review of Methods for Comparing Estrogenic Activity of Endogenous and Exogenous Chemicals in Human Milk and Infant Formula. *Environmental Health Perspectives.* 111 (8): 1020-1036.

Braden, A.W.H., N.K. Hart, and J.A. Lamberton. 1967. The oestrogenic activity and metabolism of certain isoflavones in sheep. *Aust. J. Agric. Res.*, 18:335-348.

Braga, O., G.A. Smythe, A.I. Schafer, and A.J. Feitz. 2005. Steroid estrogens in ocean sediments. *Chemosphere*, 61(6):827-833.

Brian, J.V., C.A. Harris, M. Scholze, A. Kortenkamp, P. Booy, M. Lamoree, G. Pojana, N. Jonkers, A. Marcomini, and J.P. Sumpter. 2007. Evidence of Estrogenic Mixture Effects on the Reproductive Performance of Fish. *Environmental Science & Technology*, 41(1):337-344.

Brion, F., C.R. Tyler, X. Palazzi, B. Laillet, J.M. Porcher, J. Garric, and P. Flammarion. 2004. Impacts of 17[beta]-estradiol, including environmentally relevant concentrations, on reproduction after exposure during embryo-larval-, juvenile- and adult-life stages in zebrafish (Danio rerio). *Aquatic Toxicology*, 68(3):193-217.

Brookbanks, E.O., R.A.S. Welch, and M.R. Coup. 1969a. Oestrogens in pasture and a possible relationship with mastitis. *New Zealand Veterinary Journal*, 17:159-160.

Brookbanks, E.O., R.A.H. Welch, and M.R. Coup. 1969b. Estrogens in pasture and a possible relation with mastitis. *New Zealand Veterinary Journal*, 17:159-160.

Brooks, B.W., P.K. Turner, J.K. Stanley, J.J. Weston, E.A. Glidewell, C.M. Foran, M. Slattery, T.W. La Point, and D.B. Huggett. 2003. Waterborne and sediment toxicity of fluoxetine to select organisms. *Chemosphere*, 52(1):135-142.

Campeau, R.C., L.F. Gulli, and J.F. Graves. 1996. Drug Resistance In Detroit River Gram-Negative Bacilli. *Microbios.*, 88(357):205-212.

Castiglioni, S., R. Bagnati, R. Fanelli, F. Pomati, D. Calamari, and E. Zuccato. 2006. Removal of Pharmaceuticals in Sewage Treatment Plants in Italy. *Environmental Science & Technology*, 40:357-363.

Christensen, F.M. 1998. Pharmaceuticals in the Environment-A Human Risk? *Regulatory Toxicology and Pharmacology*, 28(3):212-221.

Clara, M., B. Strenn, and N. Kreuzinger. 2003. Carbamazepine as a possible anthropogenic marker in the aquatic environment: investigations on the behaviour of Cabamazepine in wastewater treatment and during groundwater infiltration. *Water Research*, 38:947-954.

Couwenbergs, C.J. 1988. Acute effects of drinking beer or wine on the steroid hormones of healthy men. *J. Steroid Biochem*, 31:467-473.

Daughton, C.G., and T.A. Ternes. 1999. Pharmaceuticals and personal care products in the environment: agents of subtle change? *Environmental Health Perspectives*, 107(6):907-938.

De Jongh, S., L. Ose, T. Szamosi, C. Gagne, M. Lambert, R. Scott, P. Perron, D. Dobbelaere, M. Saborio, M.B. Tuohy, M. Stepanavage, A. Sapre, B. Gumbiner, M. Mercuri, A.S. Van Trotsenburg, H.D. Bakker, and J.J. Kastelein. 2002. Efficacy and safety of statin therapy in children with familial hypercholesterolemia: a randomized double-blind, placebo-controlled trial with Simvastatin. *Circulation*, 106(17):2231-2237.

Dhom, G. 1992. Epidemiology of hormone-dependent tumors. In *Endocrine-Dependent Tumors*. Edited by K. D. Voight and C. Knabbe. Raven Press, New York, NY.

Donohue, J. 2008. Personal Communication (Telephone). Edited by: United States Environmental Protection Agency, Office of Water, Office of Science and Technology, Health and Ecological Criteria Division.

Drewes, J.E., T. Heberer, and K. Reddersen. 2002. Fate of pharmaceuticals during indirect potable reuse. *Water Science and Technology*, 46(3):73-80.

Drewes, J.E., J. Hemming, S.J. Ladenburger, J. Schauer, and W. Sonzongni. 2005. An assessment of endocrine disrupting activity changes during wastewater treatment through the use of bioassays and chemical measurements. *Water Environment Research*, 77(1):12-23.

Drugs.com 2006g. Physicians Desk Reference professional listing for Risperdal [Online]. Available: <http://www.drugs.com/pro/risperdal.html>. [cited May 31, 2007]

Eaton, D.L., and C.D. Klaassen. 2001. Principles of Toxicology. In Casarett & Doull's Toxicology : The Basic Science of Poisons. *Casarett & Doull's Toxicology : The Basic Science of Poisons.*, Edited by C.D. Klaassen. New York, NY:McGraw-Hill.

Ferguson, P.L., and B.J. Brownawell. 2003. Degradation of nonylphenol ethoxylates in estuarine sediment under aerobic and anaerobic conditions. *Environmental Toxicology and Chemistry*, 22(6):1189-1199.

Fisher, J.S. 2004. Environmental anti-androgens and male reproductive health: focus on phthalates and testicular dysgenesis syndrome. *Reproduction*, 127(3):305-315.

Fitzgerald, J.E., J.L. Sanyer, J.L. Schardein, R.S. Lake, E.J. Mcguire, and F.A. De La Iglesia. 1981. Carcinogen Bioassay and Mutagenicity Studies with the Hypolipidemic Agent Gemfibrozil. *J Natl Cancer Inst.*, 67(5):1105-1116.

Fitzgerald, J.E., J.A. Petrere, and F.A. De La Iglesia. 1987. Experimental studies on reproduction with the lipid-regulating agent Gemfibrozil. *Fundam Appl Toxicol*, 8(4):454-464.

Foran, C.M., J. Weston, M. Slattery, B.W. Brooks, and D.B. Huggett. 2004. Reproductive Assessment of Japanese Medaka (Oryzias Latipes) Following a Four-Week Fluoxetine (SSRI) Exposure. *Arch Environ Contam Toxicol.*, 46(4):511-517.

Foster, W.G., C.L. Hughes, S. Chan, and L. Platt. 2002. Human developmental exposure to endocrine active compounds. *Environmental Toxicology and Pharmacology*, 12(2):75-81.

Fritsche S, Steinhart H. 1999. Occurrence of hormonally active compounds in food: a review. *Eur. Food Res. Technol.*, 209: 153-179.

Gagne, F., P. Cejka, C. Andre, R. Hausler, and C. Blaise. 2007a. Neurotoxicological effects of a primary and ozonated treated wastewater on freshwater mussels exposed to an experimental flow-through system. *Comparative Biochemistry and Physiology Part C*, 146:460-470.

Gagne, F., C. Blaise, C. Andre, C. Gagnon, and M. Salazar. 2007b. Neuroendocrine disruption and health effects in Elliptio complanata mussels exposed to aeration lagoons for wastewater treatment. *Chemosphere*, 68(4):731-743.

Garrison, A.W., J.D. Pope, and F.R. Allen. 1975. GC/MS analysis of organic compounds in domestic wastewaters: *Chem. Congr. North Am. Cont.*, p. 517-556.

Gaylor, D.W., and L.S. Gold. 1998. Regulatory Cancer Risk Assessment Based on a Quick Estimate of a Benchmark Dose Derived from the Maximum Tolerated Dose. *Regulatory Toxicology and Pharmacology*, 28:222-225.

Gold, L.S., T.H. Slone, and B.N. Ames. 2001. Natural and Synthetic Chemicals in the Diet: A Critical Analysis of Possible Cancer Hazards. *Food Safety and Food Quality*, R.E. Hester & R.M. Harrison, eds. Issues in Environmental Science and Technology 15. Cambridge, UK:Royal Society of Chemistry, 95-128.

Gold, L.S., N.B. Manley, T.H. Slone, L. Rohrbach, and G.B. Garfinkel. 2005. Supplement to the Carcinogenic Potency Database (CPDB): Results of Animal Bioassays Published in the General Literature Through 1997 and by the National Toxicology Program in 1997-1998. *Toxicological Sciences*, 85:747-808.

Gold, L. S., T. H. Slone, et al. 2006. The Carcinogenic Potency Database (CPDB). Retrieved June 1, 2007, from http://potency.berkeley.edu/.

Goldman, J.M., S.C. Laws, S.K. Balchak, R.L. Cooper, and R.J. Kavlock. 2000. Endocrine-disrupting chemicals: prepubertal exposures and effects on sexual maturation and thyroid activity in the female rat. A focus on the EDSTAC recommendations. *Crit Rev Toxicol*, 30(2):135-196.

Grace, P.B., J.I. Taylor, N.P. Botting, T. Fryatt, M.F. Oldfield, and S.A. Bingham. 2003a. Quantification of isoflavones and lignans in urine using gas chromatography/mass spectrometry. *Analytical Chimica Acta*, 315:114-121.

Grace, P.B., J.I. Taylor, N.P. Botting, T. Fryatt, M.F. Oldfield, N. Al-Maharik, and S.A. Bingham. 2003b. Quantification of isoflavones and lignans in serum using isotope dilution liquid chromatography/tandem mass spectrometry. *Rapid Commun. Mass Spectrom.*, 17:1350-1357.

Gray, L.E., and P.M.D. Foster. 2003. Significance of experimental studies for assessing adverse effects of endocrine-disrupting chemicals. *Pure and Applied Chemistry*, 75(11-12):2125-2141.

Greer M.A., Goodman G., Pleus R.C. and Greer S.E. 2002. Health Effects Assessment for Environmental Perchlorate Contamination: The Dose Response for Inhibition of Thyroidal Radioiodine Uptake in Humans. *Environm. Health Perspect.,* 110: 927-937.

Halling-Sorensen, B., S.N. Nielsen, P.F. Lanzky, F. Ingerslev, H.C.H. Lutzhoft, and S.E. Jorgensen. 1998. Occurence, fate and effects of pharmaceutical substances in the environment - a review. *Chemosphere*, 36(2):357-393.

Hamosh M. 2001. Bioactive factors in human milk. Breastfeeding 2001. Part I: The evidence for breastfeeding. *Pediatr. Clin. North Am.*, 48(1):69–86.

Harmann S, Lacorn M, Steinhart H. 1998. Natural occurrence of steroid hormones in food. *Food Chemistry.,* 62 (1): 7-20.

Hatta, T., H. Ohmori, T. Murakami, M. Takano, K. Yamashita, and M. Yasuda. 1999. Neurotoxic effects of Phenytoin on postnatal mouse brain development following neonatal administration. *Neurotoxicol. Terato.,* 21(1):21-28.

Hemming, J., W. Waller, M. Chow, N. Denslow, and B. Venables. 2001a. Assessment of the estrogenicity and toxicity of a domestic wastewater effluent flowing through a constructed wetland system using biomarkers in male fathead minnows (Pimephales promelas rafinesque, 1820). *Environmental Toxicology and Chemistry*, 20(10):2268-2275.

Hemming, J.M., W.T. Waller, M.C. Chow, N.D. Denslow, and B. Venablesc. 2001b. Assessment Of The Estrogenicity And Toxicity Of A Domestic Wastewater Effluent Flowing Through A Constructed Wetland System Using Biomarkers In Male Fathead Minnows (Pimephales Promelas Rafinesque, 1820). *Environmental Toxicology and Chemistry*. 20(10):2268–2275.

Hemming, J.D.C., M.A.E. Barman, and J.H. Standridge. 2003. Assessment of waters for estrogenic activity. Edited by Denver, CO: American Water Works Association Research Foundation (AwwaRF).

Henck, J.W., W.R. Craf, A. Black, J. Colgin, and J.A. Anderson. 1998. Pre- and postnatal toxicity of the HMG-CoA Reductase Inhibitor Atorvastatin in rats. *Toxicol Sci*, 41(1):88-99.

Hignite, C. and D. L. Azarnoff. 1977. Drugs and drug metabolites as environmental contaminants: chlorophenoxyisobutyrate and salicylic acid in sewage water effluent. *Life Sciences*. 20(2): 337-341.

Huang, C.-H., and D.L. Sedlak. 2001a. Analysis of estrogenic hormones in municipal wastewater effluent and surface water using enzyme-linked immunosorbent assay and gas chromatography/tandem mass spectrometry. *Environmental Toxicology and Chemistry*, 20(1):133-139.

Huang, C.H., and D. Sedlak. 2001b. Analysis of estrogenic hormones in municipal wastewater effluent and surface water using enzyme-linked immunosorbent assay and gas chromatography/tandem mass spectrometry. *Envrironmental Toxicology and Chemistry*(20):133-139.

Hughes, C.L., G. Liu, S. Beall, W.G. Foster, and V. Davis. 2004. Effects of genistein or soy milk during late gestation and lactation on adult uterine organization in the rat. *Experimental biology and medicine*, 229:108-117.

Ikeda, M. 1988. Multiple exposure to chemicals. *Regulatory Toxicology and Pharmacology*, 8(4):414-421.

Ingram, D., K. Sanders, M. Kolybaba, and D. Lopez. 1997. Case-control study of phyto-estrogens and breast cancer. *Lancet*, 350:990-994.

Jobling, S., M. Noylan, C.R. Tyler, G. Brighty, and J.P. Sumpter. 1998. Widespread sexual disruption in wild fish. *Environmental Science & Technology*, 32:2498-2506.

Jones, O.A.H., N. Voulvoulis, and J.N. Lester. 2004. Potential Ecological and Human Health Risks Associated With the Presence of Pharmaceutically Active Compounds in the Aquatic Environment. *Potential Ecological and Human Health Risks Associated With the Presence of Pharmaceutically Active Compounds in the Aquatic Environment*, 34:335-350.

Jontofsohn, M., M. Stoffels, A. Hartmann, G. Pfister, I. Juettner, G. Severin-Edmair, K.W. Schramm, and M. Schloter. 2002. Influence of nonylphenol on the microbial community of lake sediments in microcosms. *Science of the Total Environment*, 285:1-3.

Kavlock, R.J., G.P. Daston, C. Derosa, P. Fenner-Crisp, L.E. Gray, S. Kaattari, G. Lucier, M. Luster, M.J. Mac, C. Maczka, R. Miller, J. Moore, R. Rolland, G. Scott, D.M. Sheehan, T. Sinks, and H.A. Tilson. 1996. Research Needs for the Risk Assessment of Health and Environmental Effects of Endocrine Disruptors: A Report of the U.S. EPA-Sponsored Workshop. *Environmental Health Perspectives*, 104(4):715-740.

Kelch, W. J. and J. S. Lee. 1978. Antibiotic Resistance Patterns of Gram-Negative Bacteria Isolated From Environmental Sources. *Appl Environ Microbiol.* 36(3): 450-6.

Kelly, W.N. 2001. Potential Risks and Prevention, Part 4: Reports of Significant Adverse Drug Events. *Am J Health Syst Pharm.*, 56(15):1406-1412.

Kolpin, D.W., E.T. Furlong, M.T. Meyer, E.M. Thurman, S.D. Zaugg, L.B. Barber, and H.T. Buxton. 2002. Pharmaceuticals, hormones, and other organic waste contaminants in U.S. streams, 1999-2000: a national reconnaissance. *Environmental Science & Technology*, 36(6):1202-1211.

Kolpin, D.W., M. Skopec, M.T. Meyer, E.T. Furlong, and S.D. Zaugg. 2004. Urban contribution of pharmaceuticals and other organic wastewater contaminants to streams during differing flow conditions. *Science of the Total Environment*, 328(1-3):119-130.

Kuch, H.M., and K. Ballschmiter. 2001. Determination of endocrine-disrupting phenolic compounds and estrogens in surface and drinking water by HRGC-(NCI)-MS in the picogram per liter range. *Environmental Science & Technology*, 35(15):3201-3206.

Kuiper, G.G.J.M., J.G. Lemmen, B. Carlsson, J.C. Corton, S.H. Safe, P.T. Van Der Saag, B. Van Der Burg, and J.A. Gustafsson. 1998. Interaction of estrogenic chemicals and phytoestrogens with estrogen receptor β *Endocrinology*, 139:4252-4263.

Labadie, P., and E.M. Hill. 2007. Analysis of estrogens in river sediments by liquid chromatography-electrospray ionisation mass spectrometry: Comparison of tandem mass spectrometry and time-of-flight mass spectrometry. *Journal of Chromatography A*, 1141(2):174-181.

Lawrence Berkeley Laboratory, Berkeley, CA. 2006. The Carcinogenic Potency Database (CPDB). [Online]. Available: < http://potency.berkeley.edu/>. [cited June 1, 2007]

Loos, R., G. Hanke, and S.J. Eisenreich. 2003. Multi-component analysis of polar water pollutants using sequential solid-phase extraction followed by LC-ESI-MS. *J Environ Monit*, 5(3):384-394.

Lopez, H., L.D. Satter, and M.C. Wiltbank. 2004. Relationship between level of milk production and estrous behavior of lactating dairy cows. *Animal Reproduction Science*, 81(3-4):209-223.

Loraine, G.A., and M.E. Pettigrove. 2006. Seasonal Variations in Concentrations of Pharmaceuticals and Personal Care Products in Drinking Water and Reclaimed Wastewater in Southern California. *Environmental Science & Technology*, 40(3):687-695.

Lu, L.W., K.E. Anderson, J.J. Grady, and M. Nagamani. 2001. Effects of an isoflavone-free soy diet on ovarian hormones in premenopausal women. *The Jounral of Clinical Endocrinology & Metabolism*, 86:3045-3052.

Messripour M, Forooghi-Abary A, Dashti-Ardakani. 2002. Association between Sex Hormones in Human breast Milk and Infant Growth and Development. Arch Iranian Med., 5(3): 166-169.

Metcalfe, C.D., B.G. Koenig, D.T. Bennie, M. Servos, T.A. Ternes, and R. Hirsch. 2003a. Occurrence of neutral and acidic drugs in the effluents of Canadian sewage treatment plants. *Environmental Toxicology and Chemistry*, 22(12):2872-2880.

Metcalfe, C.D., X.-S. Miao, B.G. Koenig, and J. Struger. 2003b. Occurrence of neutral and acidic drugs in the effluents of Canadian sewage treatment plants. *Environmental Toxicology and Chemistry*, 22(12):2872-2880.

Mibu, K., J. Wada, Y. Okayasu, J. Tsumori, K. Komori, H. Tanaka, J.H. Li, M. Sasaki, and C. Sato. 2004. Distribution of estrogen, nonylphenol and its derivatives in the sediments of a shallow lake. *Water Science and Technology*, 50(5, Assessment and Control of Hazardous Substances in Water):173-179.

Morton, M.S., O. Arisaka, N. Miyake, L.D. Morgan, and B.A.J. Evans. 2002. phytoestrogen concentrations in serum from Japanese men and women over forty years of age. *Journal of Nutrition*, 132:3168-3171.

Moule, G.R., A.W.H. Braden, and D.R. Lamond. 1963. The significance of oestrogensin pasture plants in relation to animal production. *Animal Breed Abstr.*, 31:139-157.

Murkies, A., F.S. Dalais, E.M. Briganti, H.G. Burger, D.L. Healy, M.L. Wahlqvist, and S.R. Davis. 2000. Phytoestrogens and breast cancer in postmenopausal women: a case control study. *Menopause*, 7:289-296.

Musset, L. 2007. Inquiry regarding OECD Test Guidelines programme on endocrine disrupters. Edited by.

Novartis 2000. Product label for Tegretol® carbamazepine USP [Online]. Available: <www.fda.gov/cder/foi/label/2001/20234S17LBL.PDF>. [cited May 29, 2007]

Novartis Pharmaceuticals Corporation 2002. Product label for Voltaren® (Diclofenac Sodium enteric-coated tablets) [Online]. Available: <http://www.fda.gov/cder/foi/label/2006/019201s035,020142s017,020254s016lbl.pdf>. [cited May 29, 2007]

OEHHA-RCHAB. 2008. Proposition 65 Safe Harbor Levels: No Significant Risk Levels for Carcinogens and Maximum Allowable Dose Levels for Chemicals Causing Reproductive Toxicity. Edited by O. o. E. H. H. A.-R. a. C. H. A. Branch, ed. Sacramento, CA: California Environmental Protection Agency, Office of Environmental Health Hazard Assessment, Reproductive and Cancer Hazard Assessment Branch.

Ohmori, H., T. Kobayashi, and M. Yasuda. 1992. Neurotoxicity of Phenytoin administered to newborn mice on developing cerebellum. *Neurotoxicol. Teratol., 1992 May-Jun*, 14(3):159-165.

Ohmori, H., K. Yamashita, T. Hatta, S. Yamasaki, M. Kawamura, Y. Higashi, N. Yata, and M. Yasuda. 1997. Effects of low-dose Phenytoin administered to newborn mice on developing cerebellum. *Neurotoxicol. Teratol.*, 19(3):205-211.

Payne, J., C. Jones, S. Lakhani, and A. Kortenkamp. 2000. Improving the reproducibility of the MCF-7 cell proliferation assay for the detection of xenoestrogens. *Science of the Total Environment*, 248:51-62.

Pepper, M.S., S.J. Hazel, M. Hümpel, and W.D. Schleuning. 2004. 8-Prenylnaringenin, a novel phytoestrogen, inhibits angiogenesis in vitro and in vivo. *Journal of Cellular Physiology*, 199:98-107.

Monarch Pharmaceuticals 2005. Product label for Septra® tablets [Online]. Available: <http://www.fda.gov/cder/foi/label/2005/017376s056,017598s039lbl.pdf>. [cited May 29, 2007]

Phillips, I., M. Casewell, T. Cox, B. De Groot, C. Friis, R. Jones, C. Nightingale, R. Preston, and J. Waddell. 2004. Does the use of antibiotics in food animals pose a risk to human health? A critical review of published data. *Journal of Antimicrobial Chemotherapy*, 53(1):28-52.

Prasain, J.K., C.C. Wang, and S. Barnes. 2004. Serial review: flavonoids and isoflavones (phytoestrogens): absorption, metabolism, and bioactivity. *Free Radical Biology & Medicine*, 37:1324-1350.

Richards, S.M., C.J. Wilson, D.J. Johnson, D.M. Castle, M. Lam, S.A. Mabury, P.K. Sibley, and K.R. Solomon. 2004. Effects of Pharmaceutical Mixtures in Aquatic Microcosms. *Environ. Toxicol. Chem.*, 23(4):1035-1042.

Richardson, S.D. 2007. Water analysis: emerging contaminants and current issues. *Analytical Chemistry*, 79(12):4295-4324.

Richter, C.A., B.G. Timms, and F.S.V. Saal. 2004. *Prostate development: mechanisms for opposite effects of low and high doses of estrogenic chemicals*: *Endocrine Disruptors: Effects on Male and Female Reproductive Systems*. Raton, FL: CRC Press.

Robinson, A.A., J.B. Belden, and M.J. Lydy. 2005. Toxicity Of Fluoroquinolone Antibiotics To Aquatic Organisms. *Environ. Toxicol. Chem.*, 24(2):423-430.

Roche Pharmaceuticals 2006. Product label for EC-Naprosyn® (Naproxen delayed-release tablets), Naprosyn® (Naproxen tablets), Anaprox®/Anaprox® DS (Naproxen Sodium tablets), Naprosyn® (Naproxen suspension) [Online]. Available: <http://www.fda.gov/cder/foi/label/2006/020067s010,018965s013,018164s055,%20017581s105lbl.pdf>. [cited May 29, 2007]

Rodriguez-Mozaz, S., M.J.L.D. Alda, and D. Barceló. 2004. Picogram per Liter Level Determination of Estrogens in Natural Waters and Waterworks by a Fully Automated On-Line Solid-Phase Extraction-Liquid Chromatography-Electrospray Tandem Mass Spectrometry Method. *Analytical Chemistry*, 76(23):6998-7006.

Ropeik, D., and G. Gray. 2002. Risk: A Practical Guide for Deciding What's Really Safe and What's Really Dangerous in the World Around You. New York:Houghton Mifflin.

Sabaliunas, D., S. Webb, A. Hauk, M. Jacob, and W. Eckhoff. 2001. Environmental fate of Triclosan in the River Aire Basin, UK. *Water Resources*, 37(13):3145-3154.

Safe, S.H. 1995. Environmental And Dietary Estrogens And Human Health - Is There A Problem. *Environmental Health Perspectives*, 103(4):346-351.

Sahlberg, B.L., and M. Axelson. 1986. Identification And Quantitation Of Free And Conjugated Steroids In Milk From Lactating Women. *Journal Of Steroid Biochemistry And Molecular Biology*, 25(3):379-391.

Schappell, N. W. 2006. Estrogenic activity in the environment: municipal wastewater effluent, river, ponds, and wetlands. *Journal of Environmental Quality*, 35: 122-132.

Schueler, F.W. 1946. Sex-hormonal action and chemical constitution. *Science*, 103:221-223.

Schulman, L.J., E.V. Sargent, B.D. Naumann, E.C. Faria, D.G. Dolan, and J.P. Wargo. 2002. A human health risk assessment of pharmaceuticals in the aquatic environment. *Human And Ecological Risk Assessment*, 8(4): 657-680.

Schwab, B.W., E.P. Hayes, J.M. Fiori, F.J. Mastrocco, N.M. Roden, D. Cragin, R.D. Meyerhoff, V.J. D'aco, and P.D. Anderson. 2005. Human pharmaceuticals in US surface waters: A human health risk assessment. *Regulatory Toxicology and Pharmacology*, 42:296-312.

Schwarz, S., C. Kehrenberg, and T.R. Walsh. 2001. Use of antimicrobial agents in veterinary medicine and food animal production. *International Journal of Antimicrobial Agents*, 17(6):431-437.

Sedlak, D.L., J.L. Gray, and K.E. Pinkston. 2000. Understanding microcontaminants in recycled water. *Environmental Science & Technology*, 34(23):509A -515A.

Sedlak, D.L., K. Pinkston, and C.-H. Huang. 2005. Occurrence Survey of Pharmaceutically Active Compounds. Denver, CO:AwwaRF.

Servos, M.R., M. Smith, R. Mcinnis, K. Burnison, B.-H. Lee, and S.B. Peter Seto. 2007. The Presence of Selected Pharmaceuticals and Personal Care Products in Drinking Water in Ontario, Canada. *Water Quality Research Journal of Canada*, 42(2):130-137.

Setchell, K.D.R., S.P. Borriello, P. Hulme, D.N. Kirk, and M. Axelson. 1984. Nonsteroidal estrogens of dietary origin: possible roles in hormone-dependent disease. *Am. J. Clin. Nutr.*, 40:569-578.

Setchell, K.D.R., S.J. Gosselin, M.B. Welsh, J.O. Johnston, W.F. Balistreri, L.W. Kramer, B.L. Dresser, and M.J. Tarr. 1987. Dietary estrogens - a probable cause of infertility and liver disease in captive cheetahs. *Gastroenterology*, 93(2):225-233.

Setchell, K.D.R., L. Zimmer-Nechemias, J. Cai, and J.E. Heubi. 1997. Exposure of infants to phytoestrogens from soy-based infant formula. *Lancet*, 350:23-27.

Shore, L.S., M. Gurevitz, and M. Shemesh. 1993. Estrogen as an environmental pollutant. *Bulletin of Environmental Contamination and Toxicology*, 51:361-366.

Shu, X.O., F. Jin, Q. Dai, W. Wen, J.D. Potter, L.H. Kushi, Z. Ruan, Y.T. Gao, and W. Zheng. 2001. Soyfood intake during adolescence and subsequent risk of breast cancer among Chinese women. *Cancer Epidemiol. Biomarkers & Prev.*, 10:483-488.

Shutt, D.A., and A.W.H. Braden. 1968. The significance of equol in relation to the oestrogenic responses in sheep ingesting clover with a high formononetin content. *Aust. J. Agric. Res.*, 19:545-553.

Silva, F.R., and J. Palermo-Neto. 1999. Developmental, neuro and immunotoxic effects of perinatal Diazepam treatment in rats. *Immunopharmacol. Immunotoxicol.*, 21(2):247-265.

Singh, G., P.H. Driever, and J.W. Sander. 2005. Cancer Risk in People with Epilepsy: The Role of Antiepileptic Drugs. *Brain*, 128(Pt 1):7-17.

Sluczewski, A., and P. Roth. 1948. Effects of androgenic and estrogenic compounds on the experimental metamorphoses of amphibians. *Gynecology and obstetrics*, 47:164-176.

Smyth, H.F., C.S. Weil, J.S. West, and C.P. Carpenter. 1970. An Exploration of Joint Toxic Action: II. Equitoxic Versus Equivolume Mixtures. *Toxicol. Appl. Pharmacol.*, 14:498-503.

Snyder, S., T.L. Keith, D.A. Verbrugge, E. Snyder, T.S. Gross, K. Kannan, and J.P. Giesy. 1999a. Analytical methods for detection of selected estrogenic compounds in aqueous mixtures. *Envinormental Science and TEchnology*, 33(16):2814-2820.

Snyder, S.A., T.L. Keith, D.A. Verbrugge, E.M. Snyder, T.S. Gross, K. Kannan, and J.P. Giesy. 1999b. Analytical methods for detection of selected estrogenic compounds in aqueous mixtures. *Environmental Science & Technology*, 33(16):2814-2820.

Snyder, S.A., D.L. Villeneuve, E.M. Snyder, and J.P. Giesy. 2001a. Identification and Quantification of Estrogen Receptor Agonists in Wastewater Effluents. *Environmental Science & Technology*, 35(18):3620-3625.

Snyder, S.A., T.L. Keith, C.G. Naylor, C.A. Staples, and J.P. Giesy. 2001b. Identification and quantification method for nonylphenol and lower oligomer nonylphenol ethoxylates in fish tissues. *Environmental Toxicology and Chemistry*, 20(9):1870-1873.

Snyder, S.A., D.L. Villeneuve, E.M. Snyder, and J.P. Giesy. 2001c. Identification and quantification of estrogen receptor agonists in wastewater effluents. *Environmental Science & Technology*, 35(18):3620-3625.

Snyder, S.A., J. Leising, P. Westerhoff, Y. Yoon, H. Mash, and B.J. Vanderford. 2004. Biological and physical attenuation of endocrine disruptors and pharmaceuticals: implications for water reuse *Ground Water Monitoring & Remediation*, 24(2):108-118.

Snyder, S.A., E. Snyder, D. Villeneuve, K. Kurunthachalam, A. Villalobos, A. Blankenship, and J. Giesy. 2000. Instrumental and bioanalytical measures of endocrine disruptors in water. In *Analysis of Environmental Endocrine Disruptors*: *ACS Symposium Series*. Edited by L. H. KeithT. L. Jones-Lepp and L. L. Needham. Washington,DC: American Chemical Society.

Snyder, S.A., E. Wert, H. Lei, P. Westerhoff, and Y. Yoon. 2007a. Removal of EDCs and pharmaceuticals in drinking and reuse treatment processes. Edited by Denver: American Water Works Association Research Foundation.

Snyder, S.A., E.C. Wert, H.D. Lei, P. Westerhoff, and Y. Yoon. 2007b. Removal of EDCs and Pharmaceuticals in Drinking and Reuse Treatment Processes. Edited by Denver, CO: Awwa Research Foundation.

Snyder, S.A., K.L. Kelly, A.H. Grange, G.W. Sovocool, E.M. Snyder, and J.P. Giesy. 2001d. Pharmaceuticals and personal care products in the waters of Lake Mead, Nevada. In *Pharmaceuticals and personal care products in the environment*. Edited by C. Daughton and T. Jones-Lepp. Washington, D.C.: American Chemical Society.

Sole, M., D. Raldua, F. Piferrer, D. Barcelo, and C. Porte. 2003. Feminization of wild carp, Cyprinus carpio, in a polluted environment: plasma steroid hormones, gonadal morphology and xenobiotic metabolizing system. *Comparative Biochemistry And Physiology C-Toxicology & Pharmacology*, 136C(2):145-156.

Sonnenschein, C., and A.M. Soto. 1998. An updated review of environmental estrogen and androgen mimics and antagonists. *Journal of Steroid Biochemistry & Molecular Biology*, 65(1-6):143-150.

Soto, A.M., C. Sonnenschein, K. Chung, M. Fernandez, N. Olea, and F. Olea Serrano. 1995. The e-screen assay as a tool to identify estrogens: an update on estrogenic environmental pollutants. *Environmental Health Perspectives*, 103(Suppl. 7):113-122.

Stackelberg, P.E., E.T. Furlong, M.T. Meyer, S.D. Zaugg, A.K. Henderson, and D.B. Reissman. 2004. Persistence of pharmaceutical compounds and other organic wastewater contaminants in a conventional drinking-water-treatment plant. *Science of the Total Environment*, 329:99-113.

Stumm-Zollinger, E. and G. M. Fair. 1965. Biodegradation of steroid hormones. *Journal of the Water Pollution Control Federation,* 37: 1506-1510.

Tabacova, S. 2005. Mode of action: Angiotensin-converting enzyme inhibition-- developmental effects associated with exposure to ACE inhibitors. *Crit. Rev. Toxicol.*, 35(8-9):747-755.

Tabacova, S.A., and C.A. Kimmel. 2001. Enalapril: pharmacokinetic/dynamic inferences for comparative developmental toxicity. A review. *Reprod Toxicol*, 15(5):467-478.

Tabak, H.H., and R.L. Bunch. 1970. Steroid hormones as water pollutants. I. Metabolism of natural and synthetic ovulation-inhibiting hormones by microorganisms of activated sludge and primary settled sewage. *Dev. Ind. Microbiol.*, 11:367-376.

Tang, B., and N. Adams. 1980. Effects of equol on estrogen receptors and on synthesis of DNA and protein in the immature rat uterus. *J. Endocrinol.*, 85:291-297.

Ternes, T. A. 2001. Analytical methods for the determination of pharmaceuticals in aqueous environmental samples. *Trends in Analytical Chemistry*, 20(8): 419-434.

Ternes, T.A., M. Meisenheimer, D. Mcdowell, F. Sacher, H.-J. Brauch, B. Haist-Gulde, G. Preuss, U. Wilme, and N. Zulei-Seibert. 2002. Removal of pharmaceuticals during drinking water treatment. *Environmental Science & Technology*, 36(17):3855-3863.

Thompson, L.U., P. Robb, M. Serraino, and F. Cheung. 1991. Mammalian lignan production from various foods. *Nutr. Cancer*, 16:43-52.

Timm, G. 2007. Personal Communication. Email Message on September 29, 2007. Edited by Washington, DC: U.S. Environmental Protection Agency, Office of Science Coordination Policy.

Tixier, C., H.P. Singer, S. Oellers, and S.R. Mueller. 2003. Occurrence and Fate of Carbamazepine, Clofibric Acid, Diclofenac, Ibuprofen, Ketoprofen, and Naproxen in Surface Waters. *Environmental Science & Technology*, 37(6):1061-1068.

Trenholm, R.A., B.J. Vanderford, J.C. Holady, D.J. Rexing, and S.A. Snyder. 2006. Broad range analysis of endocrine disruptors and pharmaceuticals using gas chromatography and liquid chromatography tandem mass spectroscopy. *Chemosphere*, 65:1990-1998.

Tsuda, H., A. Naito, C.K. Kim, K. Fukamachi, H. Nomoto, and M.A. Moore. 2003. Carcinogenesis and its modification by environmental endocrine disruptors: in vivo experimental and epidemiological findings. *Jpn. J. Clin. Oncol.*, 33:259-270.

Turnidge, J. 2004. Antibiotic use in animals--prejudices, perceptions and realities. *Journal of Antimicrobial Chemotherapy*, 53(1):26-27.

Valentin-Blasini, L., B.C. Blount, S.P. Caudill, and L.L. Needham. 2003. Urinary and serum concentrations of seven phytoestrogens in a human reference population subset. *Journal of Exposure Analsis and Environmental Epidemiology*, 13:276-282.

Vanderford, B.J., R.A. Pearson, D.J. Rexing, and S.A. Snyder. 2003. Analysis of endocrine disruptors, pharmaceuticals, and personal care products in water using liquid chromatography/tandem mass spectrometry. *Analytical Chemistry*, 75(22):6265-6274.

Vanderford, B. J. and S. A. Snyder. 2006. Analysis of pharmaceuticals in water by isotope dilution liquid chromatography/tandem mass spectrometry. *Environmental Science & Technology*, 40: 7312-7320.

Velle, W. 1963. Metabolism of Estrogenic Hormones in Domestic Animals. *General and Comparative Endocrinology*, 3:621-635.

Versteegh, J.F.M., A.A.M. Stolker, W. Niesing, and J.J.A. Muller. 2003. Geneesmiddelen In Drinkwater En Drinkwaterbronnen {Pharmaceuticals In Drinking Water And Drinking Water Sources. In Dutch}.RIVM-report nr. 703719004/703712003 RIVM. Bilthoven, Netherlands.

Vieno, N., T. Tuhkanen, and L. Kronberg. 2006. Removal of pharmaceuticals in drinking water treatment: effect of chemical coagulation. *Environmental Technology*, 27:183-192.

Vieno, N.M., T. Tuhkanen, and L. Kronberg. 2005. Seasonal variation in the occurrence of pharmaceuticals in effluents from a sewage treatment plant and in the recipient water. *Environmental Science & Technology*, 39(21):8220-8226.

Vom Saal, F.S., and C.L. Hughes. 2005. An extensive new literature concerning low-dose effects of bisphenol A shows the need for a new risk assessment. *Environmental Health Perspectives*, 113:926-933.

Voutsa, D., P. Hartmann, C. Schaffner, and W. Giger. 2006. Benzotriazoles, alkylphenols and bisphenol A in municipal wastewaters and in the Glatt River, Switzerland. *Environmental Science and Pollution Research International*, 13(5):333-341.

Waters, A., and J. Knowler. 1982. Effect of a lignan (HPMF) on RNA synthesis in the rat uterus. *J. Reprod. Fertil.*, 66:379-381.

Webb, S., T. Ternes, M. Gibert, and K. Olejniczak. 2003. Indirect human exposure to pharmaceuticals via drinking water. *Toxicology Letters*, 142(3):157-267.

Weigel, S., U. Berger, E. Jensen, R. Kallenborn, H. Thoresen, and H. Huhnerfuss. 2004. Determination of selected pharmaceuticals and caffeine in sewage and seawater from Tromso/Norway with emphasis on ibuprofen and its metabolites. *Chemosphere*, 56(6):583-592.

Westerhoff, P., Y. Yoon, S. Snyder, and E. Wert. 2005. Fate of endocrine-disruptor, pharmaceutical, and personal care product chemicals during simulated drinking water treatment processes *Environmental Science & Technology*, 39(17):6649-6663.

Wolford S. T. and Argoudelis C. J. 1979. Measurement of estrogens in cow's milk, human milk, and dairy products. *J. Dairy Sci.*, 62: 1458-1463.

World Health Organization, I.P.O.C.S.W.I. 2002. Global Assessment of the State-of-the-Science of Endocrine Disruptors, In *World Health Organization Technical Report Series*, Edited by Geneva, Switzerland: World Health Organization.

Wu, A.H., R.G. Ziegler, P.L. Horn-Ross, A.M. Nomura, D.W. West, L.N. Kolonel, J.F. Rosenthal, R.N. Hoover, and M.C. Pike. 1996. Tofu and risk of breast cancer in Asian-Americans. *Cancer Epidemiol. Biomarkers & Prev.*, 5:901-906.

Zheng, W., Q. Dai, L.J. Custer, X.O. Shu, W.Q. Wen, F. Jin, and A.A. Franke. 1999. Urinary excretion of isoflavonoids and the risk of breast cancer. *Cancer Epidemiology, Biomarkers & Prevention*, 8:35-40.

ABBREVIATIONS AND ACRONYMS

AAD	American Academy of Dermatology
ACE	angiotensin-converting enzyme
ACGIH	American Conference of Governmental Industrial Hygienists
ACN	acetonitrile
ACS	American Cancer Society
ADI	acceptable daily intake
AERS	Adverse Event Reporting System
ATSDR	Agency for Toxic Substances and Disease Registry
AWWA	American Water Works Association
AwwaRF	Awwa Research Foundation
BHA	butylated hydroxyl anisole
BHT	butylated hydroxyl toluene
CA DHS	California Department of Health Services
°C	Celsius
d	day
DCM	dichloromethane
DEET	N,N-diethyl toluamide
DI	deionized water
DWEL	drinking water equivalent level
DWTP	drinking water treatment plant
EDC	endocrine disrupting chemical
EEq	estradiol equivalents
FDA	Food and Drug Administration
GAC	granular activated carbon
GC	gas chromatography
HLB	hydrophilic/lipophilic balance
IARC	International Agency for Research on Cancer
IIHS	Insurance Institute for Highway Safety
kg	kilogram
L	liter
LC	liquid chromatography
LOAEL	lowest observed adverse effect level

M	molarity
MBR	membrane bioreactor
MCL	maximum contaminant level
MDL	method detection limit
mg	milligram
mL	milliliter
MOA	mechanism of action
M.O.S.	margin of safety
MRL	maximum residue level
MS/MS	tandem mass spectrometry
MTBE	methyl tertiary butyl ether
MTD	maximum tolerated dose
µL	microliter
ng/L	nanograms per liter
NOAEL	no observed adverse effect level
NSAID	non-steroidal anti-inflammatory drug
NTP	National Toxicology Program
OTC	over-the-counter
PCP	personal care products
PhAC	pharmaceutically active compound
PPCP	pharmaceuticals and personal care products
QAQC	quality assurance/quality control
RL	reporting limit
RO	reverse osmosis
SF	slope factor
SPE	solid phase extraction
SSRI	selective serotonin reuptake inhibitor
STPE	sewage treatment plant effluent
TCA	trichloroacetic acid
TCEP	tris(chloroethyl) phosphate
TCPP	tris(1,3-dichloroisopropyl) phosphate
UF	uncertainty factor
US DOT	United States Department of Transportation
US EPA	United States Environmental Protection Agency
US FDA	United States Food and Drug Administration
UV	ultraviolet

v/v	volume per volume
WHO	World Health Organization
WRF	WateReuse Foundation
WWTP	wastewater treatment plant

IWA Publishing's authorised EU representative for General Product Safety Regulations is Diane D'Arras, 15 rue Duret, 75116 Paris, France, e-mail: safety@iwap.co.uk.

Printed and bound by CPI Group (UK) Ltd, Croydon, CR0 4YY
18/09/2026
02218938-0008